Imagining Air

Imagining Air

Cultural Axiology and the Politics of Invisibility

edited by
TATIANA KONRAD

UNIVERSITY
of
EXETER
PRESS

First published in 2023 by
University of Exeter Press
Reed Hall, Streatham Drive
Exeter EX4 4QR, UK
www.exeterpress.co.uk

A CIP catalogue record for this book is available from the British Library

https://doi.org/10.47788/GMTY9723

Published with the support of the Austrian Science Fund (FWF): PUB 1019-P

FWF Austrian
Science Fund

ISBN 978-1-80413-117-6 Hbk
ISBN 978-1-80413-118-3 Pbk
ISBN 978-1-80413-119-0 ePub
ISBN 978-1-80413-120-6 PDF

Cover image: Jemima Wyman
Plume 20: Smoke from burning property during riots caused by the non-guilty verdict of
police officers in the beating of Rodney King, Los Angeles, USA, 29 April 1992 (Black [B])
(x2); Smoke from burning tyres during Arab Spring, Manama, Bahrain, 28 February 2012
(B); Smoke from burning property during Euromaidan protests, Kyiv, Ukraine, 22 January
2014 (B with rain)...
2022, 450 × 530 cm, handcut digital photos. Photograph: Ed Mumford

Typeset in Adobe Caslon Pro by S4Carlisle Publishing Services, Chennai, India

Contents

Contributors

Siobhan Carroll is an Associate Professor of English at the University of Delaware with a specialization in British Romantic Literature. A writer as well as a scholar of speculative fiction, her work explores the historical interactions of empire, science, and the environment. Her first book, *An Empire of Air and Water: Uncolonizable Space in the British Imagination, 1750–1850* (2015), was the runner-up for the First Book Prize of the British Association for Romantic Studies. She is currently working on a new project on the nineteenth-century Anthropocene and the problem of agency.

Jeff Diamanti is Assistant Professor of Environmental Humanities at the University of Amsterdam and is crosslisted between Literary & Cultural Analysis and Philosophy. His first monograph, *Climate and Capital in the Age of Petroleum*, was published in 2021.

Corey Dzenko is an Associate Professor of Art History in the Department of Art and Design at Monmouth University (New Jersey, USA). She focuses her research on contemporary art and photography in terms of the politics of identity. She was a Visiting Fellow in Art History at the University of Nottingham and a co-editor of *Contemporary Citizenship, Art, and Visual Culture: Making and Being Made* (2018). She has also published with *Burlington Contemporary*, *Men and Masculinities*, *Chasqui: Revisita de Literatura Latinoamericana*, and *Afterimage: The Journal of Media Arts and Cultural Criticism*.

Clare Hickman is Reader in Environmental and Medical History at Newcastle University. Her recent Wellcome Trust funded Fellowship, "The Garden as a Laboratory," merged the history of medicine, health and science, with that of the landscape and environment and has resulted in her latest book, *The Doctor's Garden: Medicine, Science, and Horticulture in Britain* (2021). She currently leads the Wellcome Trust-funded *MedEnv: Intersections in Medical and Environmental Humanities* network and the AHRC-funded *Unlocking Landscapes Network: History, Culture and Sensory Diversity in Landscape Use and Decision Making*. She is also Co-Investigator on the AHRC standard grant, *In All Our Footsteps: Tracking, Mapping and Experiencing Rights of Way in Post-War Britain* and the NERC-funded *Connected Treescapes* project.

Tatiana Konrad is a postdoctoral researcher in the Department of English and American Studies, University of Vienna, Austria, the principal investigator of "Air and Environmental Health in the (Post-)COVID-19 World," and the editor of the "Environment, Health, and Well-being" book series at Michigan State University Press. She holds a PhD in American Studies from the University of Marburg, Germany. She was a Visiting Fellow at the University of Chicago (2022), a Visiting Researcher at the Forest History Society (2019), an Ebeling Fellow at the American Antiquarian Society (2018), and a Visiting Scholar at the University of South Alabama (2016). She is the author of *Docu-Fictions of War: U.S. Interventionism in Film and Literature* (2019), the editor of *Plastics, Environment, Culture and the Politics of Waste* (2023), *Cold War II: Hollywood's Renewed Obsession with Russia* (2020), and *Transportation and the Culture of Climate Change: Accelerating Ride to Global Crisis* (2020), and a co-editor of *Cultures of War in Graphic Novels: Violence, Trauma, and Memory* (2018).

Jayne Lewis is Professor of English at the University of California, Irvine. She is the editor of the anthology *Religion in Enlightenment England* (2016) and the author of *Air's Appearance: Literary Atmosphere in English Fiction, 1660–1794* (2012). She has published numerous articles and books on Anglophone literature and culture in the Enlightenment and beyond. Her current work explores figures and fantasies of transpersonal embodiment as these integrate the forgotten literacies of an ostensible past with the postliterate media forms of an imagined present and the specter of an illegible and possibly nonoccurring future.

Chantelle Mitchell is an independent scholar and artistic researcher. Her research interests across the environmental humanities include extraction, temporality, and affect. She has published with *Green Letters, e-flux, art+ Australia, Performance Philosophy, On_Culture,* and *un Magazine.* She holds a Bachelor's degree in Art History and Philosophy from the University of Western Australia, and a Masters in Curatorship from the University of Melbourne. She was a project assistant for "Air and Environmental Health in the (Post-)COVID-19 World" at the University of Vienna. Chantelle maintains a collaborative practice with Jaxon Waterhouse (Australia), which has seen them present at numerous Australian and international conferences, and deliver exhibitions across university and independent art spaces.

Christian Riegel is Professor of English and Medical and Health Humanities at Campion College at the University of Regina. He has published widely on literature and health, including the books *Writing Grief: Margaret Laurence and the Work of Mourning* (2003), *Response to Death: The Literary Work of Mourning* (2005), as well as on Canadian writing, including *A Sense of Place: Regionalism in Canadian and American Writing* (1998) and *Twenty-First*

Century Canadian Writers (2007). He has recently co-edited the volume *Health Humanities in Application* (2023).

Arthur Rose is a Senior Research Fellow in English at the University of Exeter in the UK, where he is affiliated to the Wellcome Trust Collaborative Award "Shame and Medicine." His most recent books are *Asbestos—The Last Modernist Object* (2022), and, with Fred Cooper and Luna Dolezal, *COVID-19 and Shame: Political Emotions and Public Health in the UK* (2023).

Gordon M. Sayre is a scholar of colonial American history and literature and Professor of English and Folklore at the University of Oregon. He is participating faculty in Environmental Studies and teaches an undergraduate course entitled Car Cultures that examines the history of the car industry and the environmental issues caused by automobility. He is also the author of "The Humanity of the Car: Automobility, Agency, and Autonomy" published in the journal *Cultural Critique* in 2020.

Savannah Schaufler is a project assistant for "Air and Environmental Health in the (Post-)COVID-19 World" at the University of Vienna and a PhD Candidate at the Doctoral School of Ecology and Evolution. During her studies at the University of Vienna, from which she graduated with honors in Evolutionary Anthropology, she participated in several inter- and trans-disciplinary projects at the intersection of cultural, human, and biological sciences. In addition, she is finishing her bachelor's degree in Cultural and Social Anthropology and has already published in peer-reviewed literature.

Introduction

Toward a Cultural Axiology of Air

Tatiana Konrad, Chantelle Mitchell, and Savannah Schaufler

> Can man live elsewhere than air? Neither in earth, nor in fire, nor in water
> is any habitation possible for him. No other element can for him take the
> place of place. … But this element, irreducibly constitutive of the whole,
> compels neither the faculty of perception nor that of knowledge to recog-
> nize it. Always there, it allows itself to be forgotten.
>
> Luce Irigaray, *The Forgetting of Air in Martin Heidegger*[1]

A poorly designed bottle sits atop a mantelpiece, contents slowly leaking
into the surrounding environment. Not toxic, but affective, this bottle contains
air collected from the Irish countryside—captured, commodified, and transported across the globe to lonely "expats" separated from families amid the
shuttering of global borders in response to the COVID-19 pandemic.[2] This
bottled air, while a novelty, was purchased in service of a very specific purpose,
that being connectivity and a brief moment of immersion. Air is a consistent
biological necessity, indispensable to human and more-than-human beings,
to life and survival. However, air is materially and ideologically complex; at
once an environmental and scientific concern amid contemporary climate
devastation, central medium for understandings of the world, and register of
entanglement, relation, and well-being. The novelty air, selling out during an
unfolding global pandemic, is a register of air's complexity, at once local and
global, present and absent, affective and material. Attention to air across
contemporary scholarship reveals that air not only carries us through space
and time, but is also fundamental to being and living.

Materially, air is a mixture of gases in the Earth's atmosphere, consisting
mainly of nitrogen and oxygen.[3] However, emissions, combustion, other
chemical substances, and biological admixtures can modify the composition
of air.[4] Biological processes are entangled with air, including respiration,

Tatiana Konrad, Chantelle Mitchell, and Savannah Schaufler, "Introduction: Toward a Cultural Axiology of Air"
in: *Imagining Air: Cultural Axiology and the Politics of Invisibility.* University of Exeter Press (2023). © Tatiana
Konrad, Chantelle Mitchell, and Savannah Schaufler. DOI: 10.47788/HWDT3673

which involves the active and passive exchange of oxygen and carbon dioxide (CO_2) within the blood, the organism, and the surrounding environment.[5] During respiration, oxygen is brought into the body through the inhalation of air and transported to cells. In turn, CO_2 is produced by these processes and released into the surrounding air when exhaled.[6] Aerosols contained in air can be absorbed through inhalation; tiny particles, invisible to the human eye, suspended in the atmosphere can enter the body through this biological process.[7] Breathing organisms need air in order to keep their cells alive. However, air can also be a barrier to health, flourishing, and well-being as a result of disease, pollution, and biopolitical impacts. In particular, the COVID-19 pandemic presents an immediate example of air's entanglement with everyday life. This pandemic demonstrates air's boundaryless nature, and its reach—a manifestation of the material fact that air itself can be dangerous, a container and vector for viruses, particulate matter, and other pollutants.

Despite the rigor with which scientific apprehensions of air seek to delineate categorical boundaries, air's complexity challenges neat frames. In recognition of the imprecision of a term such as "the air," professor of literature Eva Horn turns toward common patterns of speech, intermixed with scientific fact, to illuminate air's productive, conceptual airiness.[8] She notes that air itself is defined as the Earth's atmosphere, which entails the recognition of air's planetary reach. Further, material and temporal becomings of air manifest as weather and as climate, an ever-changing constitution of materials.[9] All the while, as David Macauley, environmental studies theorist, acknowledges, "with our heads immersed in the thickness of the atmosphere or our lungs and limbs engaged with the swirling winds, we repeatedly breathe, think and dream in the regions of the air"—at once within and without.[10] While definitions of air might appear conflated and complicated by compositional multiplicities, what is common across all approaches—scientific, semantic, and otherwise—is air's ubiquity as lived environment.

Air is united with the body through breath, a bodily process foundational to being and life, and inextricably tied to frames of illness and wellness, flourishing, and existence. Peter Sloterdijk notes that breathing is an "implicit condition of bodily existence," which surrounds us and enables our being.[11] As an embodied function, breathing unites the human body with the more-than-human world, a corporeal intra-action through which the interimplications of being as part of environments and ecologies can be evidenced. This being—human corporeality—is identified as inseparable from nature, which new materialist theorists Stacy Alaimo and Susan Hekman position as "material and discursive, natural and cultural, biological and textual" relations, manifestations of "trans-corporeality."[12] Breath as a bodily function, in which each inhalation registers direct contact with the external world, sees the alveoli, pleural cavity, the branches of the lungs, as sites of exchange—with air at once both inside and outside. In attending to the breathing body as "literal contact zone," registers of the world can be marked through social, political, material, and ecological occurrences, and as such, air can be considered a kind of ideal filled with thoughts, memories, and images.[13]

In the context of a politics of breathing, transdisciplinary researcher Marijn Nieuwenhuis claims that breath "reveals a history and a politics in itself. It is already infused with memories, chemicals, and other things of the past."[14] This is emphasized by feminist and cultural studies philosopher Magdalena Górska through her research on feminist politics of breathing and vulnerability, for she encourages thinking through breath as an "inspir[ation to] rethink the relation of embodiment, subjectivity, and environment," recognizing that through breathing we "breathe each other," an "other" that includes the human and more-than-human.[15] She engages breath through a non-universalizing and politicized understanding of human bodies as acting agents of intersectional politics. In doing so, Górska's work offers an "anthropo-situated" yet anti-anthropocentric understanding of everyday bodily and affective life practices as political affairs.[16] Within this framework, Górska posits breathing as a transformative act, more than the exchange of oxygen and CO_2, but one that articulates ethics, politics, and power relations. Górska considers this in relation to panic attacks, a rupture in steady breathing that includes feelings of breathlessness, exhaustion, and anxiety, tied to biopolitics, necropolitics, and normative notions of worth and value.[17]

Medical humanities scholar Jane Macnaughton focuses on the "symptom" in relation to breath and breathlessness, as well as attempting to qualify air's invisibility with reference to breath's inextricability and interactivity with life and being.[18] Using a critical medical humanities approach posed by William Viney, Felicity Callard, and Angela Woods, Macnaughton discusses the concepts of "lived body" and "social body."[19] She explores the relationships between breath, bodies, and the world, illuminating breath's perceived invisibility despite its centrality to many forms of life on Earth. Referring to the human body and "picking breath out from its surroundings," Macnaughton describes how breath can be sensed, heard, smelled, tasted, and seen. In this way, she highlights breath's often overlooked role in underscoring social connections and relationships.[20]

In tracing a hermeneutics of breath across Western philosophy, from Anaximenes through René Descartes, across Martin Heidegger and into Derridean frames, David Kleinberg-Levin argues for a return to a recognition of breathing as an "organ of being."[21] Indeed, contemporary turns toward philosophies of air, so-called respiratory philosophies, acknowledge the challenges made by Luce Irigaray to histories of Western thought that displace the centrality of air and breath, for thought. Irigaray identifies this *forgetting* as having created a void, "by using up the air for telling without ever telling of air itself."[22] It was precisely this forgetting that drew conceptual artist Michael Asher to air as an artistic medium, noting that "[a]ir seems to be the most obvious yet the most overlooked material that we come in everyday contact with."[23] Yet it was not visibility that Asher sought to enable through his installation practice, but rather encounter and presence through engagements with dematerialization. By manipulating fans and air-conditioning units to blow almost imperceptible drafts of air onto the passing bodies of gallery visitors with his 1969 work for the exhibition *Anti-Illusion: Procedures/ Materials* at the Whitney Museum of American Art, Asher created an encounter, a subtle reminder of air's presence through motion.[24] In the present, air subsists between these poles of forgetting and encounter, as it becomes increasingly difficult to ignore the presence, problem, and possibility of air, particularly as it is tied to breath. As CO_2 levels continue to rise, smoke from bushfires becomes so vast that it circumnavigates the globe, and pollution obscures skylines from view, the interimplications of the breathing body, air, and the world itself become key challenges for the present and the future. In line with this crucial challenge is a marked rise in scholarship attended to matters of the aerial. As acknowledged by Robert-Jan Wille, within the environmental and geohumanities, air is a significant site for contemporary research, challenging the monopoly of "land-based" scholarship in view of attention to atmospheric sensibilities.[25]

In the following pages, *Imagining Air: Cultural Axiology and the Politics of Invisibility* encourages a discussion that builds upon significant contributions to aerial scholarship in recognition of the challenges of the present. This text seeks to illuminate the social, environmental, medical, and political significance of air through the perspectives of environmental and medical humanities. Air itself presents as a means of engagement, "a way of relating to the world, of engaging with others, objects, environments, and technologies."[26] Further, considering the historical determinates of social and cultural apprehensions of air, alongside shifting materialities of air amid the ongoing climate crisis, it becomes crucial to move through the complex, entangled, and foundational space of air. In revealing air's multiplicities, *Imagining Air* reconfigures aerial relationships and values across time and place, and through matter, seeking to contribute to this rich, timely, and necessary field of study.[27] In this context, this edited collection engages with the nature of air's ascribed plurality of values, as lived experiences influence and shape its perception and attribution.[28]

Airy Materialities and Manifestations

Contained within Michel Serres' *Birth of Physics* is an impassioned lament. He writes that the "sages of yesterday, of long ago, were passionately interested in the Meteora (that is, meteorology and natural forms)," but that no one reads this "Meteora" anymore.[29] In his apprehension of the historical and contemporary field of theory, he calls for attention to "Meteora" as necessary for full view of the "landscape" and as essential to the "science of today and tomorrow."[30] Originally published in 1980, Serres' text preceded the so-called material turn that was to follow in the late 1990s and early 2000s, from which an abundance of attention to matter and natural forms emerged. This new materialism marked a shift toward the recognition of matter as active, agentic, and lively, and of the co-constitutive power of ecologies, terrestrial, and otherwise.

It was the philosopher Anaximenes of Miletus (586–526 BCE) who posited that the arche, the first principle or origin of all things, was air.[31] Yet, in the centuries that have followed, air has slipped from view, that paradoxical present absence that contributes to air's forgetting in philosophical traditions. This forgetting takes a different form in the apprehensions of Mark Jackson and Maria Fannin, who, in pursuit of an alternative current of attention in geography, seek to substitute the geo- for the aero-.[32] They recognize, in the scope of a collection of approaches to this subject, that air has perceived multiplicities and can subsequently be figured as object, matter, and medium.[33] Air is materially implicit in meteorological and emergent aerographical frames but continues its diffuse characterization even in the midst of a scholarly turn that seeks to illuminate human/more-than-human relationships. As Karen Barad emphasizes, humans are not "outside observers of the world. Nor are we simply located at particular places *in* the world; rather, we are part *of* the world in its ongoing intra-activity."[34] This statement apprehends humans as participants, contributors, and co-constitutive parts of the world. Human–aerial relationships are one manifestation of this intra-action, with respiration being a means through which this relationship can be evidenced. Through breath, the human body participates materially in the external world, a participation that can be aligned with Nancy Tuana's "viscous porosity," in which viscous is understood as being the inseparability of life from air and porosity as the presence of the external material world in the body through pores—or the lungs.[35] Such theoretical framings of human and more-than-human relationships allow for intra-active recognition, a way through which it becomes possible to attend to the materiality of air.

Theorizing in the space of climate crisis, geographer and theorist Kathryn Yussof and researcher across digital media, environments, and social research Jennifer Gabrys draw from geographical foundations a recognition of imagination as "a way of seeing, sensing, thinking, and dreaming the formation of knowledge, which creates the conditions for material interventions in and political sensibilities of the world."[36] In furthering their mobilization of the work of philosopher Richard Kearney, imagination can be distilled as the "human power to connote absence into presence," a fitting observation in recognition of air's perceived absence in theoretical, cultural, and material frames.[37] Imaginaries of air, then, present the possibility of transforming air's theoretical and perceptual absence into cultural, political, and theoretical presence. If air is seen as that "through and in which everything can appear," it can be identified as a critical zone for materialization and visibility, as much as it is read as imperceptible and diffuse.[38] In this contradictory space of presence and absence, air as a subject of inquiry has significant consequences for thinking materially and ecologically in the present. This trouble of air, to borrow the phrasing of Donna Haraway, invites productive and resonant complication, from which significant elemental and material understandings of the world as a whole, and across temporalities, continue to emerge.[39]

The pursuit of a cultural axiology of air can be derived from the theory of values.[40] Axiology itself, a mode of study emerging from but not limited to philosophy and having ramifications for cultural studies, explores the question of a value system.[41] The practice of axiology entails the study of value, examining characteristics, structures, and hierarchies that inform the construction of value.[42] As furthered by contributions to this book, cultural meanings can be understood as a particular embodiment or expression of value. In ascribing more than a merely life-sustaining value to air, conceptual frameworks for (new) ways of thinking, perceiving, and understanding air emerge.[43] Thus, the meaning of air as a material condition of being is recognized and furthered in the understanding of presence and absence, visibility and invisibility, specifically in terms of pasts, presents, and futures of air, ecologies, and narratives. In this collection, cultural axiology is turned toward questions, experiences, aesthetics, and manifestations of air, in recognition of air's continued material, environmental, biological, social, and political significance. Adopting this approach and leveraging the conceptual tools, lenses, and figurations present in humanities discourses, *Imagining Air* seeks to understand how humanity has shaped air through processes of industrialization, economic growth, climate crisis, and airborne and zoonotic diseases, with the recognition that humanity, in turn, is shaped by air. The singular but entwined chapters that comprise this edited collection draw upon, further, and refigure aerial scholarship from the nineteenth century to the present, while questioning the value of air and dissecting how that value is being undermined and transformed by anthropogenic interventions. In telling of air's multiple complexities with reference to mental and bodily well-being, pollution, consumption, scientific experimentation, technology, space, and behavior, the importance of an aerial attention is underscored and expanded by inter- and transdisciplinary modes of thinking. Pursuing a movement toward an axiology of air apprehends the diffuse and conceptual frameworks that air presents, establishing new connections with and toward air, and redefining the meaning and value of air as an element essential to all life.

The Shaping of Air

In thinking through aerial encounters, entangling breath, time, place, and environment, it becomes possible to tell "airstories"—recognitions of relationships to air and breath that allow for "collective awareness of ecopolitical intra-actions and co-constitution with air."[44] Airstories, those narratives that illuminate and refigure relationships with the air that surrounds us, emerge particularly from critical questions of environmental and racial justice. Greta Gaard untangles airstories from specific environmental and social contexts, locating these in conditions such as smog and air pollution, as well as within racialized policing and carceral systems.[45] Gaard utilizes the airstory as one method of challenging air's perceived societal, material, and experiential absence. Drawing on Buddhist and feminist New Materialist frameworks, she considers how an attentiveness to air can assist in recognition of ecological intra-actions, reliances, and interbeings. Beginning with the paucity of words in the English language to refer to air, Gaard traverses multiple "airstories" specifically in relation to smog, addressing the impacts of protest, pollution, and perception on relationships with unclean air.[46] It is possible to trace a conceptual parallel between an airstory and the Latourian "geostory."[47] This emerges in recognition of the agency of the natural world and the participants who are actants *within* and tellers *of* such a story (Bruno Latour lists novelists, scientists, engineers, activists, and citizens alongside volcanoes, tectonic plates, and the Mississippi River among these).[48] The crucial distinction here is that the geostory emerges from the "return of object and subject back to the ground."[49] However, groundedness does not need to overwrite the significance of air and atmosphere as part of this common story; instead, the tethers between ground and sky, Earth and air appear as a rich field for narrative attention amid moving toward frames of airy recognition, in both personal and planetary senses.

Planetary framings of the Earth in recognition of human-induced climate crisis entail recognition of the proliferation of Anthropocene discourses. The Anthropocene, emerging as proposed successor to the Holocene, has proliferated in contemporary discourse, linked to its use by scientists Paul Crutzen and Eugene Stoermer from the early 2000s onward. This epoch enfolds the undeniable and indelible impact of human activity upon the Earth into Earth systems frames.[50] While the Anthropocene is currently proposed as a geological epoch and not officially adopted, the reverberations of human activity into the geologic strata have significant implications for air and atmospheres. Further to this, the Anthropocene is a diffuse concept, lacking rigid, definable boundaries. For Jerry C. Zee, anthropologist of environment and politics, the Anthropocene, in all its complexities, presents as one way through which one can "[attune] to how human life is continuously decentered and reconfigured in its rapport with a planet thrown past all thresholds."[51] This approach recognizes the Anthropocene as a means by which more-than-human assemblages might appear.[52] The Anthropocene is difficult to pin down (both literally and metaphorically), and in many ways it shares characteristics with the cloud of coal-powered steam released by the engines that were ushered in during the Industrial Revolution. One emergent but continuing critique of the Anthropocene arises from the recognition of centuries of colonization and dispossession that have contributed to the rise of pollution, climate crisis, and unsustainable industrial growth. This "colonial Anthropocene" encourages a critical view of a homogeneous Anthropocene in which all humans are equally implicated, a view in which the causes of the Anthropocene are duly attributed.[53] While previous geologic epochs align with stratigraphic processes, the Anthropocene complicates strata through intervention, extraction, and pollution.

As Serpil Oppermann attests, the Anthropocene appears as "a story of scale that stretches from the deepest lithic recesses of the Earth to its unsheltered atmospheric expanses."[54] The most prominent of Anthropocene manifestations within these "atmospheric expanses" is the increase of CO_2 in the atmosphere, resulting from extraction of fossil fuels from the Earth and their consumption in energy processes.[55] Crutzen has tied the emergence of the Anthropocene to the rapid acceleration of combustion, consumption, and manufacture; placing the Industrial Revolution as the historical origin point of this shift.[56] In particular, he draws a correlation between James Watt's improvements to the steam engine in the late eighteenth century, its subsequent rapid adoption, and causal connection, to current environmental and climatological crises. Philosopher and ecologist Timothy Morton also recognizes the importance of this connection, reading Watt's engine as an "all-purpose machine," but one that ushered in the "end of the world" with its invention in April 1784, and the resultant rapid extraction and consumption of fossil fuels.[57] In following these historical plumes of steam and darkened smoke clouds into the skies, and over time, the tethers of industry to air and geology to the atmosphere become apparent. The aerial implications of lithic intervention can be contextualized through the frame of the Anthropocene. Further, not simply historicized, experiences of air pollution and harm are ongoing realities, in which air itself becomes an archive of human activity—containing within it a multitude of airstories.

The Shaping of "Human" Air: A Question of Visibility

One of the many airstories emerging across human history is that of the Industrial Revolution, which arose from dramatic shifts from agrarian to industrial economies in Great Britain from the eighteenth century and continues to reverberate into the present. The Industrial Revolution was marked by dramatic and rapid societal and economic transformation, fueled by imperialism, capitalism, and great technological change.[58] Societies previously organized around farming became reoriented toward urban infrastructures and manufacturing as a result of this rapid revolution, with more densely populated urban areas giving rise to new divisions and organization of labor.[59] With these new and accelerated developments came the rise of production and manufacture, increases in population growth, scientific development, globalization, and urbanization. A key component of this rapid acceleration, tied also to the invention of the steam engine during this period, were increases in the extraction and consumption of natural resources to fuel such developments. Tracing the emergence of the Industrial Revolution, from the transformation of agrarian societies to industrial societies, necessitates attention to economic drivers. Ever since, the process of constant progress at social, economic, and technical levels has been churning through the history of humankind, as humans have been and continue to be driving forces of social development.[60] Meanwhile, the formed industries and economies have had particularly harmful and polluting effects on the environment.[61] The emergence of both industrial and capitalist forms of organization meant that groundbreaking innovations had, and continue to have, far-reaching effects on humans at increasingly shorter intervals. Benjamin Franklin's remark that "time is money" is the unchanging maxim for industrial and economic action today.[62] German publicist, sociologist, and social psychologist Harald Welzer argues that industrialization and the resulting processes have created a system of constant work and limitless growth.[63] This tightly interwoven and mutu-ally influencing network of society, economy, technology, and science has created a dynamic whose slowing down is one of the greatest challenges facing humanity today as ecological harm and devastation increase.[64] Indeed, Justin McBrian, in the broader context of the Capitalocene (an alternative to the Anthropocene, which seeks to encompass how capital organizes and reconfigures nature), emphasizes the many destructions implicated within these structures, writing that "[t]he accumulation of capital is the accumu-lation of potential extinction," a potential that is growing amid structures of continued growth.[65]

With the ever-worsening catastrophe of environmental damage, humanity faces a great threat—climate crisis. This highlights the interconnectedness of human interaction and exploitation of natural resources. The ongoing climate crisis is the result of human activity, in particular the exploitation of natural resources and the burning of fossil fuels disrupting the balance of ecosystems.[66] Importantly, this "human activity" is not read as homogeneous and equally ascribed to all people globally. Scientific research into emissions and climate change impacts demonstrates that those least responsible for emissions are those most likely to experience the climate crisis first hand.[67] In this complex crisis landscape, climate-related waves of migration, the accelerating loss of species and habitats, and projected future societal problems, mass unemploy-ment, global conflicts, and future pandemics, are unfolding.[68] In this context, Ivan Illich, an Austrian-American author, philosopher, and theologian, writes:

> Society can be destroyed when further growth of mass production renders the milieu hostile, when it extinguishes the free use of the natural abilities of society's members, when it isolates people from each other and locks them into a manmade shell, when it undermines the texture of community by promoting extreme social polarization and splintering specialization, or when cancerous acceleration enforces social change at a rate that rules out legal, cultural, and political precedents as formal guidelines to present behaviour.[69]

Illich sought to highlight the consequences of technical progress and resultant societal alienation. His approaches and theses are a plea for the rejection of the economic norm, criticizing the twentieth-century belief that prosperity, market, consumption, and freedom are inseparable. According to Illich, it is a matter of designing a "multidimensional balance of human life" and counteracting the increasing "commodification of values."[70] In his opinion, the creation of endless chains of needs was the prerequisite for the emergence of consumer society with all its consequential effects.[71]

A significant consequential effect of unbridled consumerism and extensive growth is environmental harm, with pollutants being one expression of this. Pollution, understood as contamination—specifically environmental contamination—is a vast category of cause, contingency, and effect.[72] However, fundamentally, pollutants can be understood as some matter introduced to the environment (i.e., materials, particles, biological admixtures, or microbes) that leads to environmental degradation and destabilization.[73] Additionally, pollution might itself have natural causes, but connotations of pollution imply anthropogenic origins. Air pollution, then, is these forms of contamination transposed into aerial and atmospheric contexts. However, this simple recognition belies the complexity of air pollution, particularly in frames of power, politics, and ideology.

The foundational work of geographer and discard studies scholar Max Liboiron interrogates the relationship between or equivalence of pollution and colonization. As Liboiron acknowledges, pollution is an expression of colonialism, a contemporary problem with origins in the colonial presumption of access to and occupation of Land.[74] As identified in the connections between geostories and airstories, "land," as in the stable footing of the Earth, and air, are intimately connected—with the reach of colonial structures extending across all manner of apprehensions of place. Liboiron identifies in a continuance of patterns of structural power that pollution, and its mitigation, is predicated on unfettered access to Indigenous Land for studies, infrastructures, observation, waste management, and the like.[75] One example Liboiron employs to identify the deep tendrils of colonization as rooted in the heart of pollution is in relation to the incineration of plastic waste as a means of addressing plastic pollution, often occurring in countries far separated from the (generally Western) sources of plastic consumption.[76] These activities release pollutants and toxins into the air, contaminating communities and causing environmental and personal harm. However, such plastic incineration initiatives (as suggested by US environmental nongovernmental organization the Ocean Conservancy in 2015, who additionally wrongly attributed the bulk of plastic pollution to five countries across Asia) have at their core colonial understandings predicated on the invasion and occupation of Land.[77] Understanding pollution as deeply tied to these structures of inequality and power (additionally, resistance as seen through the successful efforts of GAIA, the Global Alliance for Incinerator Alternatives in the Philippines) reveals a profound network of relationships that produce, disperse, and amplify pollution, particularly for people and places marginalized by colonial structures of power and ideology.[78] These systems are not limited to land-based or oceanic forms of pollution, not simply that which is marked on a geographical map, but extend to air pollution of the sites above these geographical demarcations as well.

In a simultaneously localized and global reflection upon the past, present, and future amid climate emergency, Swedish human ecologist, journalist, and author Andreas Malm writes in his book *Fossil Capital* that "[t]he fossil economy was born when that fire began to be fed by the material fuel of fossil energy."[79] This material fuel is one of the foundational contributors to air pollution and climate decline in the contemporary period. Looking at the historical roots of this problem, England in particular was one of the largest contributors to emissions in the nineteenth century, specifically owing to fossil fuel consumption.[80] This consumption fuels capital growth, and, as energy humanities scholars Imre Szeman and Dominic Boyer note, energy consumption has devastating ecological consequences, particularly once "energy" is brought into industrialized processes globally.[81] Szeman is particularly attentive to the sociocultural consequences of global reliance on fossil fuels, namely oil, in terms of consumption and commodification. Mobilizing "petrocultures" as a means of encompassing the seep of oil into all aspects of social, material, political, and cultural structures in contemporary society, Szeman identifies the scope of global contemporary dependence upon energy consumption.[82] This positive feedback loop, triggered by industrialization, is reflected in expanding global economies, population growth, increasing production and consumption, and consequent increase in emissions. Recognizing that air pollution has multiple histories, causes, and symptoms, one of the biggest challenges for the present is the rise in greenhouse gas emissions, one of the most troubling being CO_2.

The presence of CO_2 in the atmosphere is causally linked to a warming climate, a link identified in historical climate data with the industrial activities of the nineteenth century tied to the progressive heating of the Earth.[83] The knowledge of the Earth's warming, a consequence of increasing CO_2 levels, is not a new discovery, with Swedish Nobel laureate Svante Arrhenius calculating, as early as 1896, that Earth's temperature would increase if CO_2 concentration in the atmosphere doubled from its historical levels.[84] Concentrations of CO_2 have been increasing since preindustrial times, primarily because of emissions and combustion of fossil fuels, deforestation, urbanization, and technologization, leading to the largest contributions to global warming and environmental degradation.[85] For more than 2 million years, CO_2 concentrations have never been as high as in 2019.[86] According to an Intergovernmental Panel on Climate Change (IPCC) report, global surface temperatures are likely to rise by more than 1.5 C by the end of the twenty-first century unless emissions are drastically reduced.[87]

Historian of Britain William M. Cavert describes the smoke plagued histories of London from the seventeenth century in his book *The Smoke of London: Energy and Environment in the Early Modern City*, illuminating the long histories of air pollution tied to industry and cities. London represented an important economic center for the workforce of the British Empire, with Cavert's London defined by a "smoky life," with houses "blackened with the unmerciful smoke of coal-fires."[88] Through this literal description of historical

London, Cavert gives insights into the manifestations and continuing presence of dirty air above the city, tracing its impacts upon the formations of
the city and resultant effects upon human populations. This exemplifies the
invisible visibility of air pollution resulting from human interference with
nature. With this (in)visibility of air pollution and unconscious thinking
that air is "*just* air," there is a general failure to recognize air's presence and
centrality to life, particularly amidst increasing anthropogenic air
pollution.[89]

Sensing air as a material condition, the many threats and challenges posed
by air to life itself become apparent. Burning forests and fossil fuels produce
smoke, which can be perceived as a white, gray, or black "cloud" depending
on its intensity.[90] Smoke is not the only atmospheric contaminant clouding
the skies and contributing in multiple ways to various compositions of air
pollution. Mixing with fog, smoke can produce smog, which usually contains
a high concentration of exhaust gases and dust particles in the ambient air.[91]
In particular, the main components of smog include pollutants such as sulfur
dioxide (SO_2), nitrogen dioxide (NO_2), and ozone (O_3).[92] These compositions,
resulting from human activities, have detrimental effects not only on human
and more-than-human bodies, but also on broader environments and ecologies. Smoke and smog make visible air's materiality, and have tangible,
physical, and biological impacts upon breathing bodies. The transformative
conditions of smoke and smog colored experiences of London, being captured
in art and literature of the time. In a scathing reflection upon the perceived
bleakness and misery of London, poet William Blake decried the consequences of industrialization in the representation of a soot-covered, darkened
landscape in his 1794 poem "London."[93] Recognizing "[m]arks of weakness,
marks of woe" in the faces and countenance of those he passes walking the
Thames, Blake ties these states to the blackening of landmarks and institutions
across the city by coal smoke and soot, while workers, soldiers and children
are weighted by the pressures of industry and society.[94] The smog-covered
city of London lends itself specifically to the Impressionist approaches of
Claude Monet, depicting lingering fog over significant landmarks, with *Houses
of Parliament (Effect of Fog)*, *Waterloo Bridge, Effect of Fog*, and *Houses of the
Parliament, Effect of Sunlight in the Fog* capturing the affective resonance of
smoke, fog, and smog across cityscapes in aesthetically beautiful registers.[95]

It is against the backdrop of air's entanglement across industry, ecology,
and health that it becomes possible to attend to the political, social, and
structural implications of air and air pollution as an airstory tethered to
questions of human and more-than-human justice. In the foundational work
of environmental humanities scholar Rob Nixon, instances of ecological harm
and pollution are positioned as a form of violence enacted upon specific
peoples.[96] This violence, positioned as "slow violence," presents a theoretical
framework through which it becomes possible to identify currents within
contemporary climatological realities. This violence illuminates how "unequal
power relations … are 'inscribed' in the air" and are subsequently marked

across landscapes, airscapes, and bodies (both human and more-than-human).[97] The slowness that Nixon speaks to is a cumulative violence, "a violence that occurs gradually and out of sight, a violence of delayed destruction that is dispersed across time and space, an attritional violence that is typically not viewed as violence at all."[98] The notion of dispersal and accretion lends itself to aerial connotations, with Nixon's characterization of slow violence as occurring "out of site" aligned with the nonvisible qualities of air and atmosphere. Significantly, "slow violence" invokes social, political, economic, and racial marginalization as key determinants of the experience of such harm. Nixon's case studies, which include considerations of nuclear fallout in the Marshall Islands following atmospheric nuclear weapons testing, or the long-term 1984 Union Carbide radiological gas leak at Bhopal, India, serve not only to highlight the lasting effects of particular forms of environmental harm, but also to consider the ways in which such narratives are told and represented.[99] For Nixon, addressing "slow violence," which has particular manifestations in the Global South, "requires … that we plot and give figurative shape to formless threats whose fatal repercussions are dispersed across space and time."[100] In apprehending the formlessness of air and its contaminants through place, experience, and matter, the telling of airstories allows for forms of attention, which illuminate structural manifestations of ecological harm and its reverberations.

In his considerations of contemporary encounters with air pollution, ecological harm, and illness in the context of contemporary cities, cultural anthropologist Timothy Choy establishes a "poetics of honghei (ambient air)" in the context of air pollution in China.[101] Taking an anthropological approach to the subject, Choy illustrates the dangers of polluted air through Beijing's extreme air pollution—dubbed the "airpocalypse" by media outlets in 2013 and 2014.[102] Toxins present in the air can impact the body in many ways, with air pollution itself leading to 4.2 million deaths globally per year, at a minimum.[103] Lived experience of air pollution informs Choy's approach, as he tells of leaving Beijing with his partner owing to recurrent and persistent sinus infections caused by the heavy pollution.[104] With a visual and private glimpse into life in Beijing, Choy demonstrates the impossibility of separating the self from air, particularly toxic air. This is extended into affective frames through his concept of the "poetics of air" that seeks to understand, feel, and express the meanings of air by incorporating social, geographic, ecological, medical, and political perspectives.[105] In enacting a "poetics of air," it becomes possible to attend to the embodied, co-constitutive, and emergent facts of air as central to human existence, clean or toxic.[106] An example of the extensive effects of air pollution on the human body, besides impairing lung function and increasing the prevalence of lung cancer, are changes in DNA methylation, in other words, the DNA molecule itself.[107] Several studies have supported the theory that environmental influences or exposures to toxic substances and pollutants can affect epigenetic developmental programming, most notably during the developmental stages of early life.[108]

Drawing attention to the problem of air pollution in 2015, Chinese air purifier company Xiao Zhu projected imagery of crying children onto the smoke emanating from factories.[109] Appearing almost like an art installation, this advertisement highlighted air pollution as a significant contemporary issue, not only within Chinese cities, but globally.[110] Confronted with the impending reality of cityscapes rendered in smog-filled environments presents not only a frightening glimpse of the future, but also holds a mirror to the reality of life for those living in megacities. It is within these unique urban environment that incomprehensible amounts of pollutant matter are suspended in the air, filtered into the lungs of the populace.[111] Attempts to render air pollution in visual media is not a new technique, with renowned geographer Peter Adey tracing this in the work of impressionist painters, but further, as highlighted by art-science scholar Caterina Albano.[112] Reading the marble gas-mask sculptures of Ai Wei Wei in the tradition of the memento-mori, or interrogating Hiwa K's critique and meditation on tear gas in his powerful film *This Lemon Tastes of Apple* (2011), Albano ties visual culture to the complexities of breath and air in the contemporary.[113] Albano rereads the durational work *Public Smog* (2004–) by environmental artist Amy Balkin in the context of environmental art practices, identifying the work as a continuation of such traditions.[114] *Public Smog* is a complex work, but one comprising actions toward the creation of an atmospheric park—with the artist purchasing emissions offsets or seeking UNESCO World Heritage status for the Earth's atmosphere.

Leveraging the affective capacities of artistic engagements with aerial-environmental crises, there is the possibility of greater societal recognition of air pollution as a worsening problem—a global problem. Indeed, it was in 2010 that the Texas Environmental Law Center brought forward a legal challenge to the State of Texas with regard to air pollution present and future, arguing that the state had a fiduciary duty to reduce emissions.[115] For most, the gradual increase in air pollution may not be immediately evident in everyday life. However, shifts in behavior in 2020 and 2021 began to paint a clear picture of the possibilities for cleaner skies. In 2020 and 2021, a reduction in air pollution amid the COVID-19 pandemic as a result of worldwide lockdowns, social distancing mandates, and resultant reductions in travel, traffic, and business activity found its way into the news cycle. Reports from Punjab, India, showed the snow-topped peaks of the Himalayas visible for the first time in decades, while blue skies in Los Angeles, USA, preceded news articles stating that the city somewhat unbelievably had some of the cleanest air in the world following state-wide lockdowns.[116] While this state of clarity was short lived, reconfigurations of human behavior were demonstrably linked for a moment to tangible reductions in air pollution.[117] Following these plumes of coal-marked steam from engines into the atmosphere, where we now encounter ever-increasing CO_2 levels, rising temperatures, and a depleted ozone layer, reveals the interconnectedness of earth systems and the influence of human activity upon their health and vitality.

The interconnectedness of human and ecological health is amplified by the experience of the COVID-19 pandemic, a global disaster that has revealed the negative effects of the prevailing imbalance between humans and the environment.[118] In recent years, the re-emergence of infectious diseases of zoonotic origin has been a major worldwide threat as well as an economic burden.[119] This threat can be tied to the intrusion of humanity into ecosystems and nature, disturbing their balance.[120] These imbalances can cause "zoonotic spillovers," with humans becoming hosts for infectious diseases.[121] Owing to these intrusions, which themselves have roots in the changes impelled by the Industrial Revolution (urbanization, globalization, technologization, and population growth), environments are in consistent states of change and degradation. Urbanization in particular is considered a strong driver of epidemiological transition, furthered by, among other things, increasing global tourism and growing demand for meat and other animal products.[122] As can be seen, the emergence and occurrence of zoonoses is amplified and accelerated by human intervention in a functioning biological system. Therefore, in understanding zoonotic diseases, especially airborne diseases such as COVID-19, a thorough comprehension of the multifactorial issue, including environment, human social behavior, political and economic changes, evolution, and pathogen life cycles is essential. In this sense, air can be considered polluted, impure, and infected by these invisible microbes that swirl through air, sickening humans and nonhumans.

History is replete with examples of new infectious disease outbreaks, especially those that emerged during local or global wars.[123] A well-known example is the plague, or the Black Death (1347–1353), caused by the bacterium *Yersinia pestis*. It was travel, the movement of populations, and trade goods that made the rapid and extensive spread of the disease possible in the first place.[124] Some assumptions suggest that prevailing conditions, such as the physical and environmental health of a population, play an important role in the spread of contagious diseases. Factors including hygiene, hunger, famine, vitamin deficiencies, waste, dirty water, and involuntary immigration can provide optimal conditions for the growth and spread of epidemics within conditions of war.[125] These early reimaginings of air through disease were being experienced through catastrophes, sickness, smoke, and fog lingering above. With the increasing presence of airborne diseases resulting in death, loss, and instability, air's complex materiality is demanding greater attention and recognition.

As invisible threats, zoonotic and airborne diseases are easily overlooked until they spill over into pandemic and epidemic proportions, with virologist Stephen Morse warning as early as 1993 "that the problem of emerging viruses may seem too vast and too amorphous to tackle," but they present a key challenge for the coming decades.[126] Likewise, science journalist Robin Marantz Henig addressed the question of how and why viruses arise in her 1994 book *A Dancing Matrix: How Science Confronts Emerging Viruses*. This publication, based on interviews with scientists and emerging from scientific research, explained how new viruses can be created by environmental changes, climate changes, and urbanization, and what influence exposure to human nature has on viruses. Henig presents viruses as small, but immensely challenging for humanity.[127] The scenario of a pandemic outbreak of a new infectious disease, not to be dismissed, was also demonstrated by simulation exercises in 2018 and 2019 by the Johns Hopkins Center for Health Security.[128] Shortly after this, COVID-19's spread through human populations triggered radical global shifts in behavior alongside widespread fears of contagion.[129] With these behavioral and atmospheric shifts, the understanding of air as a material presence and a disease vector has been altered. Similarly, the threat of air pollution is a challenging problem for the present—hanging over Earth's landscapes and populations not like a dark cloud but rather akin to the sword of Damocles. Such threats not only underscore air's complex materiality, but also make air apparent as an essential condition of being.

Aerial Encounters and Questions of Time, Place, and Justice

Foundational contemporary philosopher Peter Sloterdijk positions the violence emerging from attacks upon the breathing body as a defining characteristic of the twentieth century through the concept of "atmoterrorism"; a theoretical apprehension that reverberates into the present era.[130] Sloterdijk identifies "atmoterrorism" as having roots in the military tactics of World War I, being a form of "environmental warfare" in which state violence is focused upon creating conditions that "attack on the enemy's primary, ecologically dependent vital functions."[131] Drawing on the emergence of gas attacks as key military strategy and recognizing chemical warfare and toxins as part of modern military history, "atmoterrorism" emphasizes the locatedness of the breathing body amid atmospheres and ecologies upon which it is dependent. The introduction of poison gas as weapon during World War I, on April 22, 1915, at Ypres, Belgium, by German military forces forever altered the landscape of warfare, with the years of conflict following seeing 100,000 tons of chemical weapons deployed by Allied and Axis forces alike.[132] Chemical warfare deemed a "higher form of killing" by Nobel Prize-winning scientist Dr. Fritz Haber continued throughout the twentieth century, with devastating impacts (including, for example, the use of Zyklon B, Agent Orange, and DDT by a number of military forces throughout the 1900s).[133] However, as Nieuwenhuis recognizes in a crucial text exploring the colonial roots of "atmoterrorism," Sloterdijk's twentieth-century accounts fail to accommodate the widespread use of gas attacks as an extension of colonial rule and toward extermination of populations subjugated under invasion.[134] Specifically, Nieuwenhuis considers targeted use by colonial governments across the Middle East as a means of enforcing colonial occupation, control, and policing.[135] As the historical record openly attests, gas was used in colonial policing by the British in occupied Palestine in the 1930s and by French forces across occupied North Africa in the same period.[136] This specific use by the French was not new, with Nieuwenhuis positioning the extermination of populations by governmental forces in the Vendée region in the late eighteenth century and in Algeria in the late nineteenth century as central to colonial practices of occupation and genocide.[137]

The history of chemical warfare, specifically gas attacks, reveals the extent to which the control of air and airspaces is tied to power. The use of chemical warfare may have been officially prohibited by the *UN Convention on the Prohibition of Military or Any Other Hostile Use of Environmental Modification Techniques*, which came into force in 1978, but "terror from the air" shifted, altered, and changed as aerial weaponry continued to advance over the twentieth and into the twenty-first century. In perceiving this airscape, "atmo-terrorism" demonstrates the vitality of breath for the human body. Moving beyond the frame of warfare, the body can be read as a "contact zone" within which political, social, and environmental forces coalesce.[138]

Reading across the events of the 2000s, namely Occupy Wall Street, 9/11, and the Black Lives Matter (BLM) movement, Franco Bifo Berardi acknowledges breathing as political.[139] Through the poetry of Friedrich Hölderlin, Berardi arrives at a position that mirrors the recognitions of embodiment, corporeality, and breath: "the intimate texture of being is breathing."[140] However, when breath is turned against the breathing body, being and its possibilities for flourishing and existence are challenged. What happens when breath is difficult? What happens when one cannot breathe? In tracing the textures of being across breath and through air, material, discursive, and political forces can be identified in the diffuse space of air. The breathing body as a point of contact between the self and the external world reads differently across environments, impacted by material, political, and social contexts of experience, allowing us to uncover truths in proximity to environments. In the context of breath, environmental humanities scholar Jean-Thomas Tremblay also emphasizes the political resonance of the breathing body in the context of intermixing political, social, and environmental forces. Tremblay writes that the "pollution, weaponization, and monetization [of breath and air has] turned them into dispositifs of state control and violence. It's also the case that, under such conditions, breathing has become a minoritarian aesthetic and political vernacular in its own right."[141] Here, Tremblay coalesces the mobilization of artistic and political responses to contemporary complexities emerging from structures and powers that control and influence breath and breathing.

The BLM movement began as a Black-centered political movement in 2013 resulting from collective outrage following the acquittal of a police officer charged with the murder of 17-year-old Trayvon Martin, in Sanford, Florida. BLM quickly spread across the globe via social media with a hashtag (#BlackLivesMatter), in response to widespread structural racism perpetrated by those in power against Black, Indigenous, and People of Color (BIPOC) communities, with in-person protests quickly following. These protests emerged in response to acts of police violence, specifically the assassination of Eric Garner in 2014 by officers of the New York City police force. Footage of Garner's murder was widely circulated across social media alongside calls to action, and within this footage Garner was heard to repeat "I can't breathe" multiple times while placed in a chokehold. Since this incident in 2014, this phrase has been heard in numerous other assassinations of Black Americans by police forces, most notably Javier Ellis in 2019 and Manuel Ellis and George Floyd in 2020, and taken up as a collective chant in BLM protests against police brutality. If breath, that is the inhalation of air, is an essential condition of being, denying the possibility of breath is denying the possibility of being, which illuminates biopolitical harms faced by those oppressed by structural powers.[142]

The state-sanctioned violence experienced by BIPOC communities globally is not inseparable from legacies of racism and environmental harm. Leveraging Nixon's "slow violence," public health and environmental justice researchers Margaret Hicken, Lewis Miles, Solome Haile, and Michael Esposito call upon numerous studies that connect proximity to environmental harm through proximity to factories or waste dumps and disposal sites to inequality in environmental hazard exposure for predominantly Black communities in a US context.[143] Alongside the increased presence of urban heat islands for Black and Latino communities, or the greater likelihood of hazardous waste facilities in proximity to Black communities, there is evidence demonstrating that Black Americans have greater exposure to air pollution than white Americans, despite contributing less to such pollution.[144] These structures recognize structures of harm and subjugation that emerge not only from white privilege, but also continued racism enacted socially and politically. Alongside the climate crisis, scholar Anita Lam recognizes narratives of breath-taking within climates of anti-Blackness in the USA, which entangle not only police violence and unequal exposure to environmental harm, but also, through the COVID-19 pandemic, atmospheric harm.[145] For Lam, climate is a "sensitizing metaphor" within which racism can be recognized as an "atmospheric condition."[146] Further, within the COVID-19 pandemic, unequal distribution and response to harm caused by exposure to the virus manifested as an extension of Christina Sharpe's "virulent antiblackness."[147] While there is metaphorical power in speaking to racism in these terms, through narratives and imagery of climates, environments, atmospheres, and breath, these frameworks are structures through which it is possible to apprehend legacies and lived experience of colonization, enslavement, and subjugation that predominately white Western nations continue to exploit for gain. Indeed, decades prior to current protest movements, Franz Fanon identified in the context of colonization that

> [t]here is not occupation of territory, on the one hand, and independence of persons on the other. It is the country as a whole, its history, its daily pulsation that are contested, disfigured, in the hope of a final destruction. Under these conditions, the individual's breathing is an observed, an occupied breathing. It is a combat breathing.[148]

While earlier invocations of cohesive breath identify occurrences within regimes of policing or in the military industrial complex, Fanon calls attention to an atmospheric manifestation of state control—breath, that material condition of life.

The recognition of materiality and politicization of air necessitates attention to air as space. Materially, air has the characteristic of transience. This characteristic manifests in satellite imagery of smoke from catastrophic bushfires circumnavigating the globe or large-scale sandstorms moving terrestrial matter between continents. Beyond this material conception of air, it is also conceived as a location or a site, with terrestrial borders extending to demarcate airspace as belonging to a country or nation. While air has the characteristics of a commons, it is experienced differently by different populations and is tied to structures of inequality. The right to air is policed and weaponized. Beyond this, air is monitored, surveilled, and controlled by the extension of nation-states beyond terrestrial confines. While this extension of a state's border into the air is a known condition of contemporary geopolitical life (the recent events of the COVID-19 pandemic, which have seen the shutting of borders and the cessation of most air travel globally), there is no international declaration that provides a precise definition of "vertical sovereignty."[149] Air, conceived as a site or location, is complicated by interactions between material notions of transience and terrestrial notions of fixity. As aforementioned, air has the characteristics of a commons, but this commons is troubled through the demarcations of aerial space in the frame of national power, in terms of borders and control. While Matthew Longo might position a border as the "definitive marker of the political," borders, as they extend through aerial space, are not as materially or conceptually rigid as they appear.[150]

The problem of aerial borders emerges in the context of surveillance, warfare, transit, and culpability. As artist-geographer Trevor Paglen identifies, the contemporary period is marked by cartographic impulse, entangled with contemporary geography.[151] This impulse follows from a colonial lineage, which sees "spatial enclosure" via circumnavigation and occupation expanding toward aerial, atmospheric, and planetary territories.[152] Central to this specific expansion are technologies of surveillance that allow virtual exploration of the globe from a "'God's Eye' vantage-point."[153] This above-ground view has similarly been tied to the notion of an "Apollonian gaze," one that "pulls diverse life on earth into a vision of unity."[154] However, as revealed with attention to the multiplicities of air and its social, political, collective, and individual manifestations, the notion of aerial unity is troubled. Further, this particular conception of an aerial unity is one tied to specific geopolitical structures, informed by the construction of airspace through war.

Airpower, initially defined as the ability to do "something in the air," is a notion with its origins in twentieth-century warfare, enacted in traditional military frameworks of defense.[155] In the twenty-first century, the notion of airpower has expanded toward surveillance, particularly surveillance as removed from the viewing body, positioned in the framework of state-orchestrated terror. In the performance of this specific manifestation of airpower, it is no longer necessary for humans to take to air—remaining grounded while unmanned aerial vehicles, or drones, extend the gaze beyond the viewing body, across shores and into other national territories.[156] Marking an era of surveillance, alongside satellites surrounding the Earth in a steel ring, drones transform air into a militarized space, crossing borders and complicating relationships to air and aerial space.

The infiltration of air by technologies of surveillance and warfare extends also to the literal presence of technologies within air. At its most expanded, this can be read through the traces of mineral extraction measured as CO_2 parts per million, but an alternative is to recognize air as a site of a dispersed digital. Moving through air, alongside pollution, pathogen, and politics, are manifestations of this—whether Wi-Fi or cellphone data or messages sent and received continually from satellites. Like letters through the post, air is filled with messages sent and received, encoded, and recoded, dispersed and simultaneously localized, complicated by notions of visibility as tied to presence. As philosopher and curator Monika Bakke acknowledges, "[h]uman users constantly develop new ways of using air in order to emit, receive (extract) and process … we contribute to the atmosphere through the vital respiratory cycle, but unlike nonhuman users, we saturate air with electronic data, shaping the air space not only on a physical, but also on the mental level."[157] If air is viewed in this way—as container and messenger, as physical and conceptual space—it becomes possible to read messages in other aerial traces. Beyond the digital, pollution can be read as a message of ecological crisis, smog as a message of labor and industry, pathogens as a messenger of human/more-than-human proximity. These messages complicate air's (in)visibility, foreground its materiality, and invite approaches that address air as a complex site for investigation.

Organization of the Book

The chapters in this book, divided into three parts, each trace a multitude of relationships to air. This undertaking, which spans time, place, and methodology, reveals air's increasing complexity, given the ways in which worldwide experiences of air pollution, broader environmental degradation, political structures, and the COVID-19 pandemic influence human-aerial relationships. With this in mind, the aim of this edited collection is to build on widespread scholarly engagements with and about air by bringing it to the fore in a variety of specific case studies and locations. The collection therefore reflects the diverse contexts in which air can be experienced and explored by tethering transdisciplinary contributions to the development of an axiological framework of air. Drawing on historical, literary, philosophical, cultural, medical, ecological, material, and technoscientific trajectories, the authors in this edited collection examine aerial formations presented in American, Canadian, and European contexts. These chapters entangle with a number of thematic complexities—attending to pollution, technology, visual art, photography, aesthetics and embodiment, pollution, behavior, and consumerism, and wider trans-sensory environments and landscapes.

"Part I: Aerial Politics: Pollution, Consumerism, and Catastrophe" draws on various disciplinary approaches, including geohumanities, extraction studies, pollution, and environmental hazards. The chapters within this section approach air in recognition of the depth of human–environmental interactions, particularly with reference to the underlying imperialist and capitalist impacts on the formation of air across time and place. Each chapter addresses some form of technology in the shaping and influencing of aerial relationships. From historical formations of nation-states as entwined with stories of polluted air, to shifting and changing relationships to automobiles and industry, and speculative contextualizations of aerial surveillance and exploration, these chapters present contexts through which the political and material are central to aerial formations in the present.

In Chapter 1, "Fordism in Detroit, Consumerism in Los Angeles: A Brief History of Automobile Emissions Regulation and Lessons for Greenhouse Gas Pollution," Gordon M. Sayre positions Fordism, coined by Antonio Gramsci, as emblematic of mass production and mass consumption. By paying relatively high wages, Henry Ford, the eponym, indirectly stimulated demand for the products his company produced, since employees were also able to afford the produced cars. Drawing on Fordism and consumerism, Sayre analyzes landscapes of emission regulations specifically in the context of the automobile industry. Attendant to the future, Sayre also questions if consumers and activists will continue to rely on technical solutions and innovations, or if behavioral adjustments and changes will be necessary in the increasingly unstable landscape of the climate crisis. Sayre presents that, if a neo-Fordist approach were pursued instead of the current consumerist regime, then car-drivers would be encouraged to think of themselves as producers responsible for the far wider operations and actions of their cars, rather than as consumers of products.

In Chapter 2, "Dirty Air: Literary Tropes of the Canadian Nation," Christian Riegel explores the relationship between extractive industries, pollution, and nation-building in Canadian contexts, on the basis of a novel, *The Red-Headed Woman with the Black Black Heart*, by Canadian poet and novelist Birk Sproxton about the history of the 1934 mining strike in the town of Flin Flon. Basic elements of nation-building, such as transportation (railroads and highways), exploration, trade, and transformation of the natural landscape, are presented by Riegel within a historical frame that acknowledges the continuing impacts upon Canada's social and ecological presents and futures. While nation-building represents the construction of a national identity, it also involves a redefinition of populations that had been divided by colonial powers without regard to ethnic, religious, or other boundaries. In this process, a national identity was to be deliberately constructed by forming different ethnic groups into one nation, thus creating ethnically heterogeneous populations. The relationship of disease, the destruction of the environment, and the annihilation of the First Nations people of Canada are subject in this chapter. Through these processes and the literary descriptions of nation-building, railroad transportation, and mining the effects of air pollution, disease, neglect, abuse, and poverty become binding. Within this context, Riegel ties air pollution directly to the consequences of colonial and genocidal nation-building activities and reflects broader processes of ecological harm as extensions of the subjugation of First Nations people and their sovereignty.

Chantelle Mitchell, in Chapter 3, "Witnessing *Challenger*: Viewing Aerial Space through the Reverberations of Disaster," draws on the 1986 *Challenger* space shuttle disaster to study air as not simply mere background, or invisible matter, but rather connective medium and conduit, within which complex relationships to vision, verticality, and understandings of the world are revealed. Utilizing visual analysis and archival reports, alongside theoretical considerations from disaster studies, contemporary philosophy, and the environmental humanities, this chapter seeks to illuminate the *Challenger* explosion as uniquely revelatory in the context of relationships to air and atmosphere. Doing so positions the relationship between understandings of aerial space and technological control as an extension of political will and vision, arriving at a speculative rendering of geo- and aero- formations.

Comprising a number of approaches to air that address embodiment and environment through arts and cultural practice, "Part II: Air and Art in Times of Crisis" attends to depictions of social and environmental interactions through creative mediums and climate relations. The spatiality of air in artworks, literary texts, and speculative fiction is utilized to make the imagined transparent air visible to its readers. This involves reading air in the context of interiority and exteriority amid external threats to well-being, as well as affective consequences of air's contamination. Against this backdrop, the authors engage with geophysics, cultural anthropology, photography, literature, aesthetics, and curatorship. These engagements seek to emphasize the critical role of creative practice in elucidating air's myriad complexities, particularly as they relate to aesthetics, science, economics, and technology.

Siobhan Carroll's Chapter 4, "Speculative Fiction, Atmotechnic Ecology, and the Afterlife of Romantic Air," uses London-based immersive art collective Marshmallow Laser Feast's 2021 installation "We Live in an Ocean of Air" as a springboard to question the ways in which we engage with air and its transparency. Surveying fiction from the early nineteenth century to the present, Carroll argues for the special role of speculative fiction in representing the ecological threats and community-building potential of invisible air. Drawing on Peter Sloterdijk's theory of atmotechnics, she focuses on the representation of air in Sarah Pinsker's *A Song for a New Day* and Emily St. John Mandel's *Station Eleven*—two novels about airborne disease that predate the COVID-19 pandemic—in order to anticipate the socioimaginary changes forced by the present pandemic moment.

In Chapter 5, "Respiratory Realism: Elemental Intimacies between 'Carbon Black' and *Red Desert*," Jeff Diamanti uses air in the context of hydrophysical and psychosocial flows. He shows the effects that weather patterns can have on social and psychological aspects expressed through epistemic habits, critical attitudes, and aesthetic iterations. In reference to artistic abstractions and views, he discusses the consideration of these effects and how climate realism can be created. Diamanti's climate realism is an expression of how to be caught by the currents of artwork, and how we might consider embodiment in relation. His approach leverages theory and abstract approaches, while drawing upon cultural materials, including Michelangelo Antonioni's 1964 film *Red Desert*. Doing so, Diamanti demonstrates the great influence the environment has on the body, particularly in relation to embodiment and as carried by the current of the air, the seas, hydrophysical flows, and the world.

That photography and art became a point of social interaction during the pandemic and national lockdowns is demonstrated by Corey Dzenko in Chapter 6, "Rumpled Bedsheets and Online Mourning: Social Photography and the COVID-19 Pandemic—Haruka Sakaguchi's *Quarantine Diary* and Marvin Heiferman's Instagram account @whywelook." As COVID-19 spread through the air, many turned to virtual environments to avoid more dangerous "public" spaces. This chapter discusses two specific photographic projects. In examining photography's role in forms of mourning and self-care, Dzenko elucidates the social life of photography and ways in which politics of identity and place influenced the unfolding impact of this global pandemic. With her research focus on contemporary art and photography in relation to the politics of identity, she creates an intersection of contemporary art and theory, incorporating photography and new media.

Previous chapters in this book dealt with technology and pollution in space and time, as well as considerations of cultural and literary contributions to understandings and formations of air and atmosphere. "Part III: Trans-Sensory Air: Bodies and Environments" is inclusive of approaches that consider questions of health and well-being related to the body and in the context of place. These bodies include abled, disabled, human, and nonhuman bodies, and extend to the formation of space in recognition of air and air pollution. Like many contributions to this book, the chapters within Part III address COVID-19 implicitly, or explicitly, tying contemporary experiences of polluted air to historical figurations and representations. In this way, these chapters illuminate air's trans-sensory potentialities, its experience, representation, and mutual influence on culture and behavior.

Chapter 7, "Envisioning Experiments on Air and the Nonhuman," concentrates on the presence and absence of air, revealing profound insights into scientific understanding of air. The problem of speaking from the perspective of a life without air is discussed by Arthur Rose in reference to key moments in lyrical texts, novels, and short stories. He examines the techniques used by Andrew Miller, Ted Chiang, and Daisy Lafarge to represent the treatment of nonhuman lives in experimental spaces deprived of air, particularly in the context of scientific experimentation. Stripping air of what makes it ordinary for life, these aesthetic reproductions of key moments in the scientific history of air help to make sense of our relation with this most elusive of elements. Reflecting the current context of a global respiratory pandemic, Rose shows how life without air plays a fundamental role in considering that we all breathe the same air, and breath is therefore intimately connected to life, mind, and body.

Chapter 8, "The Importance of 'Open-Air' for Health: Environmental and Medical Intersections," directly relates access to clean air to notions of hygiene and prosperity. Clare Hickman considers the role of "open air" for both health and recreational reasons—timely, and necessary as it is precisely open air that has become a central location for populations in a global pandemic. In doing so, she discusses concerns during the COVID-19 pandemic regarding the importance of access to clean air between urban centers and regions, and how this is reigniting debates and tensions around inequalities and access to the land. In comparison, the use of rural air as a preventive and therapeutic medical measure against chronic diseases such as tuberculosis in the late nineteenth and early twentieth centuries is presented. It is precisely these benefits of the outdoors that represent a key element in the perception and use of the countryside by an increasingly industrialized and urbanized population.

For Jayne Lewis, air is a complex place of presence and absence, thereby enabling an understanding of what air is through a speculative framework in Chapter 9, "'The Endless Space of Air': Helen Keller's Auratic Worldbuilding." Drawing on theories and practices of auratic photography and Helen Keller's autobiographies, Jayne Lewis's text addresses the "radiant activity" of air and thus the potential of disability to illuminate trans-sensory environments in and through the body. The autobiographical and literary texts of Helen Keller are drawn upon to discuss the connection and phenomenology of air with life, the atmosphere, the environment in which we find ourselves. Air is described by Keller as an "endless space" in which seen and unseen correspondences can be noticed. Based on this contribution, Lewis demonstrates the impact that literary experiences have upon the reader. With Helen Keller's descriptions, literary and ecstatic atmospheres emerge, captivating the reader and revealing new perspectives about the composition and effects of air.

In Chapter 10, "Questions of Visibility: Aerial Relations across Society and the Environment, as Revealed by COVID-19," Savannah Schaufler addresses ripples of air pollution in relation to human behavior, by drawing on the COVID-19 pandemic to pursue an interdisciplinary approach toward illuminating air's connectivity. In this context, Schaufler considers a contemporary worldview of "forgetting," tethering air pollution to understandings of spread and ubiquity, by drawing upon pandemic imaginaries. She argues for an alignment between air pollution and pandemics, with air pollution not only transporting pathogens across bodies, borders, nations, and cultures, but also illuminating narratives and structures of exclusion.

Imagining Air takes a comprehensive approach in situating and engaging with the prevailing scholarly debates and trends around air in history, the social sciences, and the humanities that has emerged in the present, particularly through attention to health and well-being inclusive of and beyond the human. This edited collection thus seeks to contribute to rich fields of aerial scholarship in order to frame air's materiality and criticality in the context of relationships to the world. The chapters in this collection attend to broader questions of society, aesthetics, environment, and the body, considering landscapes, technologies, pasts, and futures in a nuanced way. Across these diverse approaches, which draw from philosophical, cultural, medical, and social perspectives, these chapters expand the scales within which air—between the individual and the planetary—can be understood. The intent of this collection is to broaden and deepen existing concerns about air, which have emerged as wholly necessary amid the COVID-19 pandemic. Here, air is recognized as embedded within a broad spectrum of entanglements, challenging a politics of invisibility, and reflecting upon the complex, converging, and conflicting multiplicities of air in contemporary life.

With air come connotations of immateriality, diffusion, and imperceptibility. These connotations have been challenged in previous decades amid a discipline spanning material and environmental turns. In the work of Mark Jackson and Maria Fannin, there is recognition that "[i]f we begin with the assumption that the elemental is the material as solid, then we limit drastically the scope and relevance of our conceptual ecologies."[158] Looking to turns of phrase, these limitations can be read as connotations of air—commonly denigrated as lacking power or solidity. To be *air-headed* or *airy* is read as a negative condition; it is to lack and to be indefinite—undesirable qualities in a world that is seen to privilege solid states. Vocabularies that shape understandings of the world see that which is "grounded" as stable and desirable, and "airy" as complex, diffuse, and challenging: The "invisibility of breath has not only individual, social, and political but also linguistic manifestations."[159] With air's (re)emergence amid human-environmental frames, this challenge to fixity can be read as a productive, inspiring site for research, particularly amid a contemporary condition that is seemingly undergoing continual destabilization. As Adey recognizes, "air, then, is not just of the world 'out there'; rather, it shapes all manner of expression and forms of representation—our stories, histories, thoughts, feelings and emotions are guided by it and may well be characterized by it."[160] Each contribution to this collection seeks in some way to further attention to air's value, drawing upon multiple cross-disciplinary perspectives and axiological frameworks that continue in the expansion of conceptual ecologies to inclusive material and theoretical frames.

This edited collection has emerged from the 2022 lecture series *Air: Perspectives from the Environmental and Medical Humanities* at the University of Vienna. Among the contributors to this collection are renowned authors working in the environmental and medical humanities, attending to literature, visual art, place, time, and matter from interdisciplinary perspectives, giving shape to emergent apprehensions of air beyond purely scientific frameworks. Further, as the lived experiences of the COVID-19 pandemic demonstrate, clean air and thriving ecologies are central to sustainable planetary futures. In attending to air as complex agentic matter, air becomes recognizable as an increasingly critical domain for research through the lenses of politics, culture, peoples, and place. Thus, this book is aimed at an academic and research audience, particularly in the humanities and social sciences, but considering the implication of all within aerial environments, also draws significant ties across many fields of attention.

Beyond the scientific framework, issues of environmental health are clearly visible through the climate crisis, as evidenced by a growing discourse on sustainability, clean energy, pollution, and temperature across scholarly and artistic practices. With a growing recognition of the importance of aerial, and more broadly planetary, well-being, the contributions to this edited collection present a framework toward an axiology of air through which human–aerial relationships can be traced across place and time. With the necessity of attention to environmental health and pandemics in the present period, this volume brings together diverse approaches to a uniting issue—that of thinking the air in responsive, embodied, contextual, and axiological frames across the past, into the present, and for the future.

Notes

1   Luce Irigaray, *The Forgetting of Air in Martin Heidegger* (London: The Athlone Press, 1999), 8.

2   Jane Dunford, "Buy the Smell of 'Home' with a Bottle of UK Air—Yours for £25," *The Guardian*, December 22, 2020, https://www.theguardian.com/travel/2020/dec/22/buy-the-smell-of-home-with-a-bottle-of-uk-air-yours-for-25, accessed February 1, 2023.

3   Detlev Möller, *Luft: Chemie, Physik, Biologie, Reinhaltung, Recht* (Berlin: De Gruyter, 2003), 1. https://doi.org/10.1515/9783110200225

4   Möller, *Luft*, 2–4.

5   Friedrich Anderhuber, Timm J. Filler, Franz Pera, and Elmar T. Peuker, "Innere Organe in Thorax, Abdomen Und Becken," in *Waldeyer—Anatomie Des Menschen* (Berlin: De Gruyter, 2012), 436.

6   Roger P. Smith, Frank H. Netter, and Carlos A.G. Machado, *The Netter Collection of Medical Illustrations* (Philadelphia, PA: Elsevier, 2011), 3.

7   The Center for Aerosol Impacts on Chemistry of the Environment (CAICE), "Introduction to Aerosols," The Center for Aerosol Impacts on Chemistry of the Environment (CAICE) (2020), https://caice.ucsd.edu/introduction-to-aerosols/, accessed May 5, 2022.

8   Eva Horn, "The Case of Air," *Anthropogenic Markers: Stratigraphy and Context* (2022), https://www.anthropocene-curriculum.org/anthropogenic-markers/critical-environments/contribution/the-case-of-air, accessed April 27, 2022.

9   Horn, "The Case of Air."

10 David Macauley, "The Flowering of Environmental Roots and the Four Elements in Presocratic Philosophy: From Empedocles to Deleuze and Guattari," *Worldviews* 9, no. 3 (2005): 307. https://doi.org/10.1163/156853505774841687

11 Peter Sloterdijk, *Terror from the Air* (Cambridge, MA: Semiotext(e)/Foreign Agents, 2009), 93.

12 Stacey Alaimo, "Trans-Corporeal Feminisms and the Ethical Space of Nature," in *Material Feminisms*, ed. Stacey Alaimo and Susan Hekman (Bloomington: Indiana University Press, 2008), 238.

13 Stacey Alaimo, *Bodily Natures: Science, Environment and the Material Self* (Bloomington: Indiana University Press, 2010), 2; Peter Adey, "Air's Affinities: Geopolitics, Chemical Affect and the Force of the Elemental," *Dialogues in Human Geography* 5, no. 1 (2015): 54–75. https://doi.org/10.1177/2043820614565871

14 Marijn Nieuwenhuis, "Atemwende, or How to Breathe Differently," *Dialogues in Human Geography* 5, no. 1 (2015): 91. https://doi.org/10.1177/2043820614565876

15 Magdalena Górska, "Why Breathing Is Political," *Lambda Nordica* 26, no. 1 (2021): 110. https://doi.org/10.34041/ln.v26.723

16 Górska, "Why Breathing is Political," 154.

17 Magdalena Górska, "Corpo-Affective Politics of Anxious Breathing," in *Feminist Visual Activism and the Body* (New York: Routledge, 2020), 204–05. https://doi.org/10.4324/9780429298615-14

18 Jane Macnaughton, "'Making Breath Visible': Reflections on Relations between Bodies, Breath and World in the Critical Medical Humanities," *Body & Society* 26, no. 2 (2020): 34–37. https://doi.org/10.1177/1357034X20902526

19 William Viney, Felicity Callard, and Angela Woods, "Critical Medical Humanities: Embracing Entanglement, Taking Risks," *Medical Humanities* 41, no. 1 (2015): 2–7. https://doi.org/10.1136/medhum-2015-010692; Macnaughton, "Making Breath Visible," 34–37.

20 Macnaughton, "Making Breath Visible," 34–37.

21 David Michael Kleinberg-Levin, "Logos and Psyche: A Hermeneutics of Breathing," in *Atmospheres of Breathing*, ed. Lenart Škof and Petri Berndtson (New York: SUNY Press, 2018), 14.

22 Irigaray, *Forgetting of Air*, 8.

23 Kavior Moon, "From Air to Architecture: Michael Asher's Early Air Works," in *Conceptualism and Materiality: Matters of Art and Politics*, ed. Christian Berger (Leiden, Brill, 2019), 14. https://doi.org/10.1163/9789004404649_004

24 Whitney Museum of American Art, *Anti Illusion: Procedures/Materials* (New York: Whitney Museum of American Art, 1969), 11.

25 Robert Jan Wille, "Keep Focusing on the Air: COVID-19 and the Historical Value of an Atmospheric Sensibility," *Journal for the History of Environment and Society* 5 (2021): 185. https://doi.org/10.1484/J.JHES.5.122474

26 Rebecca Oxley and Andrew Russell, "Interdisciplinary Perspectives on Breath, Body and World," *Body & Society* 26, no. 2 (2020): 3–29. https://doi.org/10.1177/1357034X20913103

27 Marijn Nieuwenhuis, "Breathing Materiality: Aerial Violence at a Time of Atmospheric Politics," *Critical Studies on Terrorism* 9, no. 3 (2016): 499–521. https://doi.org/10.1080/17539153.2016.1199420

28 Barry Smith and Alan Thomas, "Axiology," in *Routledge Encyclopedia of Philosophy* (London: Taylor and Francis, 1998). https://doi.org/10.4324/9780415249126-L120-1

29   Michel Serres, *The Birth of Physics* (Manchester: Clinamen Press, 2000), 67.

30   Serres, *Birth of Physics*, 67.

31   Stephen Connor, *The Matter of Air: Science and Art of the Ethereal* (London: Reaktion Books, 2010), 15.

32   Mark Jackson and Maria Fannin, "Letting Geography Fall Where It May— Aerographies Address the Elemental," *Environment and Planning D: Society and Space* 29, no. 3 (2011): 435–44. https://doi.org/10.1068/d2903ed

33   Jackson and Fannin, "Aerographies," 435–44.

34   Karen Barad, "Posthumanist Performativity: Toward an Understanding of How Matter Comes to Matter," *Signs: Journal of Women in Culture and Society* 28, no. 3 (2003): 828. https://doi.org/10.1086/345321

35   Nancy Tuana, "Viscous Porosity: Witnessing Katrina," in *Material Feminisms*, ed. Stacy Alaimo and Susan Hekman (Bloomington: Indiana University Press, 2008), 188–89.

36   Kathryn Yusoff and Jennifer Gabrys, "Climate Change and the Imagination," *WIREs Climate Change* 2, no. 4 (2011): 516. https://doi.org/10.1002/wcc.117

37   Richard Kearney, *Poetics of Imagining: Modern to Post-Modern* (New York: Fordham University Press, 1998), 4.

38   Irigaray, *Forgetting of Air*, 41.

39   Donna Haraway, *Staying with the Trouble* (Durham, NC: Duke University Press, 2016).

40   Antony Flew, *A Dictionary of Philosophy* (London: Palgrave Macmillan, 1979), 32.

41   Flew, *Dictionary of Philosophy*, 32.

42   Ludwig Grünberg, *The Mystery of Values: Studies in Axiology*, ed. Cornelia Grünberg and Laura Grünberg (Amsterdam: Rodopi, 2000), 4–6.

43   Samuel L. Hart, "Axiology—Theory of Values," *Philosophy and Phenomenological Research* 32, no. 1 (1971): 29. https://doi.org/10.2307/2105883

44   Greta Gaard, "(Un)storied Air, Breath & Embodiment," *ISLE: Interdisciplinary Studies in Literature and Environment* 29, no. 2 (2020): 300. https://doi.org/10.1093/isle/isaa138

45   Greta Gaard, "New Ecocriticisms: Narrative, Affective, Empirical and Mindful," *Ecozon@: European Journal of Literature, Culture and Environment* 11, no. 2 (2020): 229. https://doi.org/10.37536/ECOZONA.2020.11.2.3520

46   Greta Gaard, "(Un)storied Air," 3–21.

47   Bruno Latour, "Agency at the Time of the Anthropocene," *New Literary History* 45, no. 1 (2014), 6. https://doi.org/10.1353/nlh.2014.0003

48   Latour, "Time of the Anthropocene," 17.

49   Latour, "Time of the Anthropocene," 16.

50   Paul Crutzen and Eugene Stoermer, "The 'Anthropocene,'" *Global Challenge Newsletter* 41 (2000): 17–18, http://www.igbp.net/download/18.316f18321323470177580001401/1376383088452/NL41.pdf, accessed May 4, 2023.

51   Jerry C. Zee, *Continent in Dust: Experiments in a Chinese Weather System* (Berkeley: University of California Press, 2022), 20.

52   Zee, *Continent in Dust*, 20.

53   Macarena Gómez-Barris, "The Colonial Anthropocene: Damage, Remapping, and Resurgent Resources," *Antipode* (2019), https://antipodefoundation.org/2019/03/19/the-colonial-anthropocene/, accessed May 4, 2023.

54   Serpil Oppermann, "The Scale of the Anthropocene: Material Ecocritical Reflections," *Mosaic: An Interdisciplinary Critical Journal* 51, no. 3 (2018): 2. https://doi.org/10.1353/mos.2018.0027

55 Oppermann, "Anthropocene," 2.

56 Paul Crutzen, "The 'Anthropocene,'" in *Earth System Science in the Anthropocene*, ed. Eckart Ehlers and Thomas Krafft (Berlin: Springer, 2006), 13–18. https://doi.org/10.1007/3-540-26590-2_3

57 Timothy Morton, *Hyperobjects: Philosophy and Ecology after the End of the World* (Minneapolis: University of Minnesota Press, 2013), 4–5.

58 Gregory Clark, "The Industrial Revolution," in *Handbook of Economic Growth* 2 , ed. Philippe Aghion and Steven N. Durlauf (Amsterdam: North Holland, 2014): 218. https://doi.org/10.1016/B978-0-444-53538-2.00005-8

59 Anthony J. McMichael, "The Urban Environment and Health in a World of Increasing Globalization: Issues for Developing Countries," *Bulletin of the World Health Organization* 78, no. 9 (2000): 1117–26, https://www.ncbi.nlm.nih.gov/pmc/articles/PMC2560839/, accessed October 6, 2022.

60 Mike Lewis, "Perspectives on Eastern Cape Industrialization," *ECSECC*, no. 20 (2016): 5, https://ecsecc.org/documentrepository/informationcentre/wp-20-industrialisationscreen_76132.pdf, accessed April 13, 2022.

61 McMichael, "Urban Environment," 1117–26.

62 Benjamin Franklin, "Advice to a Young Tradesman: To My Friend A. B.," *The New England Magazine of Knowledge and Pleasure (1758–1759)* (1759): 27. https://founders.archives.gov/documents/Franklin/01-03-02-0130

63 Harald Welzer, "Aus Fremdzwang wird Selbstzwang. Wie das Wachstum in die Köpfe kam," in *Exit: Mit Links aus der Krise*, ed. Blätter für deutsche und internationale Politik (Berlin: Blätter Verlagsgesellschaft mbH, 2011), 48.

64 Dennis Meadows, Donella Meadows, Erich Zahn, and Peter Milling. *Die Grenzen des Wachstums: Bericht des Club of Rome Zur Lage Der Menschheit* (Stuttgart: Deutsche Verlags-Anstalt, 1972), xiv.

65 Justin McBrien, "Accumulating Extinction: Planetary Catastrophism in the Necrocene" in *Anthropocene or Capitalocene? Nature, History, and the Crisis of Capitalism*, ed. Jason Moore (Oakland, CA: PM Press, 2016), 116.

66 Glenn Althor, James E.M. Watson, and Richard A. Fuller, "Global Mismatch between Greenhouse Gas Emissions and the Burden of Climate Change," *Scientific Reports* 6 (2016): 1–6. https://doi.org/10.1038/srep20281

67 United Nations Conference on Trade and Development, "Smallest Footprints, Largest Impacts: Least Developed Countries Need a Just Sustainable Transition," *UNCTAD*, October 1, 2021, https://unctad.org/topic/least-developed-countries/chart-october-2021, accessed May 4, 2023.

68 Les Convivialistes. *Das Konvivialistische Manifest*, ed. Frank Adloff and Claus Leggewie (Bielefeld: transcript, 2014), 39–59; László Szombatfalvy, *Die Größten Herausforderungen Unserer Zeit* (Hamburg: Hoffman und Campe, 2011), 25.

69 Ivan Illich, *Tools of Conviviality* (Glasgow: William Collins Sons & Co., 1973), 11.

70 Illich, *Conviviality*, 10; Julia Brein, "Autonomie, Konvivialität und Gemeinschaft. UNITIERRA—ein netzwerk für autonomes lernen in Mexiko" (Master's thesis, University of Vienna, 2011), 43. https://services.phaidra.univie.ac.at/api/object/o:1274619/get

71 Illich, *Conviviality*, 91–98.

72 Frank R. Spellman, *The Science of Environmental Pollution* (Boca Raton, FL: CRC Press, 2021); Richard Fuller, Philip J Landrigan, Kalpana Balakrishnan, Glynda Bathan, Stephan Bose-O'Reilly, Michael Brauer, Jack Caravanos, Tom Chiles, Aaron

Cohen, Lilian Corra, Maureen Cropper, Greg Ferraro, Jill Hanna, David Hanrahan, Howard Hu, David Hunter, Gloria Janata, Rachael Kupka, Bruce Lanphear, Maureen Lichtveld, Keith Martin, Adetoun Mustapha, Ernesto Sanchez-Triana, Karti Sandilya, Laura Schaefli, Joseph Shaw, Jessica Seddon, William Suk, Martha María Téllez-Rojo, Chonghuai Yan, "Pollution and Health: A Progress Update," *The Lancet Planetary Health* 6, no. 6 (2022): e535–47, https://doi.org/10.1016/S2542-5196(22)00090-0; Diane Boudreau, Melissa McDaniel, Erin Sprout, and Andrew Turgeon, "Pollution," *National Geographic*, December 14, 2022, https://education.nationalgeographic.org/resource/pollution, accessed January 30, 2023.

73 Spellman, *Environmental Pollution*; Fuller et al., "Pollution and Health"; Boudreau et al., "Pollution."

74 Max Liboiron, *Pollution is Colonialism* (Durham, NC: Duke University Press, 2021), 5–7.

75 Liboiron, *Pollution is Colonialism*; see also Glen Sean Coulthard, *Red Skin, White Masks: Rejecting the Colonial Politics of Recognition* (Minneapolis: University of Minnesota Press, 2014), 7–13.

76 Liboiron, *Pollution is Colonialism*, 11, 76.

77 Liboiron, *Pollution is Colonialism*, 11, 76.

78 GAIA, "GAIA and #BREAKFREEFROMPLASTIC Members Respond to Ocean Conservancy's Apology," *No Burn*, July 15, 2022, https://www.no-burn.org/gaia-bffp-respond-to-oc-apology/, accessed January 31, 2023.

79 Andreas Malm, *Fossil Capital: The Rise of the Steam Power and the Roots of Global Warming* (London: Verso, 2016), 24.

80 Malm, *Fossil Capital*, 27.

81 Imre Szeman and Dominic Boyer, "Introduction: On the Energy Humanities," in *Energy Humanities: An Anthology* (Baltimore, MD: Johns Hopkins University Press, 2017), 12.

82 Imre Szeman, "Energy, Climate and the Classroom: A Letter," in *Teaching Climate Change in the Humanities*, ed. Stephen Siperstein, Shane Hall, and Stephanie LeMenager (London: Routledge, 2017), 46.

83 IPCC, "Summary for Policymakers," in *Climate Change 2021: The Physical Science Basis. Contribution of Working Group I to the Sixth Assessment Report of the Intergovernmental Panel on Climate Change*, ed. V. Masson-Delmotte, P. Zhai, A. Pirani, S.L. Connors, C. Péan, S. Berger, N. Caud, Y. Chen, L. Goldfarb, M.I. Gomis, M. Huang, K. Leitzell, E. Lonnoy, J.B.R. Matthews, T.K. Maycock, T. Waterfield, O. Yelekçi, R. Yu, and B. Zhou (Cambridge: Cambridge University Press, 2021), 5.

84 Svante Arrhenius, "On the Influence of Carbonic Acid in the Air upon the Temperature of the Ground," *Philosophical Magazine and Journal of Science* 41, (1896): 268. https://doi.org/10.1080/14786449608620846

85 Shabana Parveen, Abdul Qayyum Khan, and Sohail Farooq, "The Causal Nexus of Urbanization, Industrialization, Economic Growth and Environmental Degradation: Evidence from Pakistan," *Review of Economics and Development Studies* 5, no. 4 (2019): 721–30, https://doi.org/10.26710/reads.v5i4.883; IPCC, "Summary for Policymakers," in *Climate Change 2014: Impacts, Adaptation, and Vulnerability. Part A: Global and Sectoral Aspects. Contribution of Working Group II to the Fifth Assessment Report of the Intergovernmental Panel on Climate Change*, ed. C.B. Field, V.R. Barros, D.J. Dokken, K.J. Mach, M.D. Mastrandrea, T.E. Bilir, M. Chatterjee, K.L. Ebi,

Y.O. Estrada, R.C. Genova, B. Girma, E.S. Kissel, A.N. Levy, S. MacCracken, P.R. Mastrandrea, and L.L. White (Cambridge: Cambridge University Press, 2014), 1–32.

86 IPCC, 2021, 8.

87 IPCC, 2021, 8–23; The United States National Oceanic and Atmospheric Administration reports an average temperature increase of 0.08 C per decade since 1850, an increase that has more than doubled in recent years, with records of 0.18 C from the 1980s onwards (NOAA National Centers for Environmental Information, "State of the Climate: Monthly Global Climate Report for Annual 2020," January 2021, https://www.ncei.noaa.gov/access/monitoring/monthly-report/global/202013), accessed July 7, 2022.

88 William M. Cavert, *The Smoke of London: Energy and Environment in the Early Modern City* (Cambridge: Cambridge University Press, 2016), xiv.

89 Anneleen Kenis and Maarten Loopmans, "Just Air? Spatial Injustice and the Politicisation of Air Pollution," *Environment and Planning C: Politics and Space* 40, no. 3 (2022): 1–9. https://doi.org/10.1177/23996544221094144; Malm, Fossil Capital, 20–24.

90 John C. Wells, "Smoke," in *Longman Dictionary of Contemporary English* (Edinburgh: Pearson International, 2014).

91 Fatima Jabeen, Zulfiqar Ali, and Amina Maharjan, "Assessing Health Impacts of Winter Smog in Lahore for Exposed Occupational Groups," *Atmosphere* 12, no. 11 (2021): 1. https://doi.org/10.3390/atmos12111532

92 Jabeen et al., "Assessing Health Impacts," 1.

93 Blake, William, "London," *Songs of Innocence and of Experience*, 1795, relief etching with watercolor, 4.4 × 2.7 in. (11.1 × 6.9 cm), Yale Center for British Art, Yale University.

94 Blake, "London."

95 Claude Monet, *Houses of Parliament*, 1903–1904, oil on canvas, 32 × 36.3 in. (81.3 × 92.4 cm), Metropolitan Museum, New York, https://www.metmuseum.org/art/collection/search/437128, accessed May 5, 2022; Claude Monet, *Waterloo Bridge. Effect of Fog*, 1903, oil on canvas, 25.7 × 39.7 in. (65.3 × 101 cm), The State Hermitage Museum, St. Petersburg, Russia, https://www.hermitagemuseum.org/wps/portal/hermitage/digital-collection/01.+Paintings/28490, accessed May 5, 2022; Claude Monet, *Houses of Parliament, Effect of Sunlight in the Fog*, 1904, 31.4 × 35.8 in. (80 × 91 cm), Private Collection.

96 Rob Nixon, *Slow Violence and the Environmentalism of the Poor* (Cambridge, MA: Harvard University Press, 2011).

97 Raymond Bryant, "Power, Knowledge and Political Ecology in the Third World: A Review," *Progress in Physical Geography* 22, no. 1 (1998), 89. https://doi.org/10.1177/030913339802200104

98 Nixon, *Slow Violence*, 2.

99 Nixon, *Slow Violence*, 7.

100 Nixon, *Slow Violence*, 10.

101 Timothy Choy, "Air's Substantiations," in *Lively Capital: Biotechnologies, Ethics, and Governance in Global Markets*, ed. Sunder Rajan Kaushik (Durham, NC: Duke University Press, 2012), 128.

102 Oliver Wainwright, "Inside Beijing's Airpocalypse—A City Made 'Almost Uninhabitable' by Pollution," *The Guardian*, December 16, 2014, https://www.

theguardian.com/cities/2014/dec/16/beijing-airpocalypse-city-almost-uninhabitable-pollution-china, accessed May 3, 2022.

103 World Health Organization (WHO), "Ambient Air Pollution," World Health Organization (WHO), 2022, https://www.who.int/teams/environment-climate-change-and-health/air-quality-and-health/ambient-air-pollution, accessed May 3, 2022.

104 Choy, "Air's Substantiations," 122–52.

105 Choy, "Air's Substantiations," 122–52; Sasha Engelmann, "Toward a Poetics of Air: Sequencing and Surfacing Breath," *Transactions of the Institute of British Geographers* 40, no. 3 (2015): 430–44. https://doi.org/10.1111/tran.12084

106 Engelmann, "Toward a Poetics of Air," 430–44.

107 Robert Feil and Mario F. Fraga, "Epigenetics and the Environment: Emerging Patterns and Implications," *Nature Reviews Genetics* 13, no. 2 (2012): 97–109, https://doi.org/10.1038/nrg3142; Keith M. Godfrey, Karen A. Lillycrop, Graham C. Burdge, Peter D. Gluckman, and Mark A. Hanson, "Epigenetic Mechanisms and the Mismatch Concept of the Developmental Origins of Health and Disease," *Pediatric Research* 61, no. 5, Part 2 (2007): 5R–10R. https://doi.org/10.1203/pdr.0b013e318045bedb

108 Feil and Fraga, "Epigenetics and the Environment," 97–109; Godfrey et al., "Epigenetic Mechanisms," 5R–10R.

109 Vicky Wong, "Chinese Company Uses Smoke Art to Warn of Air Pollution Dangers," *Hong Kong Free Press*, June 29, 2015, https://hongkongfp.com/2015/06/29/chinese-company-uses-smoke-art-to-warn-of-air-pollution-dangers/, accessed April 26, 2022.

110 Wong, "Smoke Art to Warn of Air Pollution."

111 Peter Adey, "Air/Atmospheres of the Megacity," *Theory, Culture & Society* 30, no. 7–8 (2013): 292. https://doi.org/10.1177/0263276413501541

112 Adey, "Air/Atmospheres of the Megacity," 292–93.

113 Caterina Albano, *Out of Breath: Vulnerability of Air in Contemporary Art* (Minneapolis: University of Minnesota Press, 2022), 57–58.

114 Albano, *Out of Breath*, 19–20.

115 Texas Commission on Environmental Quality, Appellant, v Angela Bonser-Lain; Karin Ascott, as next friend on behalf of T.V.H. and A.V.H., minor children; and Brigid Shea, as next friend on behalf of E. B. U., a minor child, Appellees, No. 03-12-00555-CV. Court of Appeals of Texas, Third District, Austin. Filed: July 23, 2014.

116 Sarah Al-Arshani, "Before and After Photos Show How Stay-at-Home Orders Helped Los Angeles Significantly Reduce its Notorious Smog," *Business Insider*, April 8, 2020, https://www.businessinsider.com/photos-stay-at-home-order-reduced-los-angeles-notorious-smog-2020-4, accessed April 25, 2022.

117 Khurram Shehzad, Muddassar Sarfraz, and Syed Ghulam Meran Shah, "The Impact of COVID-19 as a Necessary Evil on Air Pollution in India during the Lockdown," *Environmental Pollution* 266 (2020): 1–5. https://doi.org/10.1016/j.envpol.2020.115080

118 Joseph Alfred Cotruvo, A. Dufour, G. Rees, J. Bartram, R. Carr, D.O. Cliver, G.F. Craun, R. Frayer, V.P.J. Gannon, and World Health Organization (WHO), "Waterborne Zoonoses: Identification, Causes, and Control," *IWA Publishing* (2004).

119 Antonio Cascio, M. Bosilkovski, A.J. Rodriguez-Morales, and G. Pappas, "The Socio-Ecology of Zoonotic Infections," *Clinical Microbiology and Infection* 17, no. 3 (2011): 336–42. https://doi.org/10.1111/j.1469-0691.2010.03451.x

120 Cascio et al., "Socio-Ecology of Zoonotic Infections," 336–42.

121 Cascio et al., "Socio-Ecology of Zoonotic Infections," 339.

122 Susanne E. Bauer, Ulas Im, Keren Mezuman, and Chloe Y. Gao, "Desert Dust, Industrialization and Agricultural Fires: Health Impacts of Outdoor Air Pollution," *Journal of Geophysical Research: Atmospheres* 124 , no. 7 (2019): 4104–20. https://doi.org/10.1029/2018JD029336

123 Examples include the outbreaks of typhoid fever during Napoleon's attack on Russia in 1812 and during the Spanish-American War in 1898, and the high prevalence and contagion of influenza in 1916 to 1918 during World War I. Malaria also caused major casualties during World War II and the Vietnam War.

124 Sarah Douglas, "The Black Death and Nation-State Wars of the 14th Century: Environment, Epigenetics, Excess, and Expiation, 1346–1450," in *Epidemics and War: The Impact of Disease on Major Conflicts in History*, ed. Rebecca M. Seaman (Santa Barbara, CA: ABC-CLIO, LLC, 2018), 54.

125 Douglas, "The Black Death," 5276; Cascio et al., "Socio-Ecology of Zoonotic Infections," 336–42.

126 Stephan S. Morse, *Emerging Viruses* (New York: Oxford University Press, 1993), viii.

127 Robin Marantz Henig, *A Dancing Matrix: How Science Confronts Emerging Viruses* (New York: Vintage Books, 1994).

128 Uri Friedman, "We Were Warned," *The Atlantic*, March 18, 2020, https://www.theatlantic.com/politics/archive/2020/03/pandemic-coronavirus-united-states-trump-cdc/608215/, accessed October 6, 2022.

129 Friedman, "We Were Warned."

130 Sloterdijk, *Terror from the Air*.

131 Sloterdijk, *Terror from the Air*, 15.

132 J. K. Miettinen, "The Chemical Arsenal: The Time to Defuse Is Now," *Bulletin of the Atomic Scientists* 30, no. 7 (1974): 38. https://doi.org/10.1080/00963402.1974.11458140

133 Diana Preston, *A Higher Form of Killing: Six Weeks in World War I That Forever Changed the Nature of Warfare* (London: Bloomsbury Publishing, 2015).

134 Marijn Nieuwenhuis, "Atmospheric Governance: Gassing as Law for the Protection and Killing of Life," *Environment and Planning D: Society and Space* 36, no. 1 (2017): 78–95. https://doi.org/10.1177/0263775817729378

135 Nieuwenhuis, "Atmospheric Governance," 84.

136 Martin Thomas, "'Paying the Butcher's Bill': Policing British Colonial Protest after 1918," *Crime, Histoire & Sociétés / Crime, History & Societies* 15, no. 2 (2011): 71–72. https://doi.org/10.4000/chs.1288

137 Nieuwenhuis, "Atmospheric Governance," 84–85.

138 Nieuwenhuis, "Breathing Materiality," 499–521.

139 Franco Bifo Berardi, *Breathing: Chaos and Poetry* (Los Angeles, CA: Semiotext(e), 2018).

140 Berardi, *Breathing*, 16.

141 Jean-Thomas Tremblay, *Breathing Aesthetics* (Durham, NC: Duke University Press, 2022), 31.

142 Ashon T. Crawley, *Blackpentecostal Breath: The Aesthetics of Possibility* (New York: Fordham University Press, 2016).

143 Margret T. Hicken, Lewis Miles, Solome Haile, and Michael Esposito, "Linking History to Contemporary State-Sanctioned Slow Violence through Cultural and Structural Racism," *The ANNALS of the American Academy of Political and Social Science* 694, no. 1 (2021): 51. https://doi.org/10.1177/00027162211005690

144 Bill M. Jesdale, Rachel Morello-Frosch, and Lara Cushing, "The Racial/Ethnic Distribution of Heat Risk-Related Land Cover in Relation to Residential Segregation," *Environmental Health Perspectives* 121, no. 7 (2013): 811–17, https://doi.org/10.1289/ehp.1205919; Christopher W. Tessum. Joshua S. Apte, Andrew L. Goodkind, Nicholas Z. Muller, Kimberley A. Mullins, David A. Paolella, Stephen Polasky, Nathaniel P. Springer, Sumil K. Thakrar, Julian D. Marshall, and Jason D. Hill, "Inequity in Consumption of Goods and Services Adds to Racial-Ethnic Disparities in Air Pollution Exposure," *Proceedings of the National Academy of Sciences* 116, no. 13 (2019): 6001–06. https://doi.org/10.1073/pnas.1818859116

145 Anita Lam, "Criminal Anthroposcenes 2.0: Race, Racism, and Breath-Taking Violence in the Time of COVID," *Crime, Media, Culture* 19, no. 1 (2022): 1–17. https://doi.org/10.21428/cb6ab371.470eecd1

146 Lam, "Criminal Anthroposcenes."

147 Christina Sharpe, *In the Wake: On Blackness and Being* (Durham, NC: Duke University Press, 2016), 109.

148 Frantz Fanon, *A Dying Colonialism* (New York: Grove Press, 1994), 65.

149 Dean N. Reinhardt, "The Vertical Limit of State Sovereignty," *Journal of Air Law and Commerce* 72, no. 1 (2007): 66–135. https://scholar.smu.edu/jalc/vol72/iss1/4, accessed April 25, 2022.

150 Matthew Longo, *The Politics of Borders: Sovereignty, Security, and the Citizen after 9/11* (Cambridge: Cambridge University Press, 2017), xii.

151 Trevor Paglen, "Experimental Geography: From Cultural Production to the Production of Space," in *Critical Landscapes; Art, Space, Politics*, ed. Emily Eliza Scott and Kirsten Swenson (Oakland: University of California Press, 2015), 41.

152 Elizabeth DeLoughrey, "Satellite Planetarity and the Ends of the Earth," *Public Culture* 26, no. 2 (2014): 257–80. https://doi.org/10.1215/08992363-2392057

153 Paglen, "Experimental Geography," 41.

154 Denis Cosgrove, *Apollo's Eye: A Cartographic Genealogy of the Earth in the Western Imagination* (Baltimore, MD: Johns Hopkins University Press, 2001), xi.

155 William Mitchell, *Winged Defense: The Development and Possibilities of Modern Air Power—Economic and Military* (New York: G.P. Putnam's Sons, 1925), xii.

156 Afxentis Afxentiou, "A History of Drones: Moral(e) Bombing and State Terrorism," *Critical Studies on Terrorism* 11, no. 2 (2018): 301–20. https://doi.org/10.1080/1753 9153.2018.1456719

157 Monika Bakke, *Going Aerial: Air, Art, Architecture* (Maastricht: Jan van Eyck Academie, 2007), 10.

158 Jackson and Fannin, "Aerographies," 435–44.

159 Oxley and Russell, "Interdisciplinary Perspectives," 3–29.

160 Peter Adey, *Air: Nature and Culture* (London: Reaktion Books, 2014), 9.

PART I

AERIAL POLITICS: POLLUTION, CONSUMERISM, AND CATASTROPHE

1

Fordism in Detroit, Consumerism in Los Angeles: A Brief History of Automobile Emissions Regulation and Lessons for Greenhouse Gas Pollution

Gordon M. Sayre

The problem of global climate change caused by greenhouse gases (GHG), chiefly carbon dioxide (CO_2) and methane, has changed conceptions of air pollution. Poor air quality was long considered a regional problem caused by sources such as forest fires, coal-burning industries, and—the topic of this chapter—emissions from internal combustion engines. That older form of pollution still exists, of course, and yet today CO_2 emissions are an additional threat, measured on a global scale and expected to cause long-term global consequences. Reducing GHG emissions involves ameliorating a global commons by balancing shared effort and sacrifice among hundreds of sovereign nations and millions of emission sources, with many controversies over historical inequities and relative risks. This chapter offers a history of air pollution caused by automobiles in the United States in the twentieth century, as a case study of the challenges of regulating both industrial and consumer sources of pollution. It links two major economic and industrial ideologies of the twentieth century, Fordism and consumerism, to the changing forms of pollution caused by cars and trucks, and the methods of controlling and regulating this pollution. I argue that air pollution is not an unintended or unwelcome by-product of the car industry. It is in fact closely bound up with the ideologies of modern automobility and political economy, and I focus on cars because they are the largest cause of air pollution attributable directly to individual consumers. The regulatory policies that have, with notable

Gordon M. Sayre, "Fordism in Detroit, Consumerism in Los Angeles: A Brief History of Automobile Emissions Regulation and Lessons for Greenhouse Gas Pollution" in: *Imagining Air: Cultural Axiology and the Politics of Invisibility*. University of Exeter Press (2023). © Gordon M. Sayre. DOI: 10.47788/TCVP5497 success, reduced air pollution by cars and trucks in the United States and the European Union have been based in consumerist ideology, but to better address GHG pollution and thereby address climate change, a return to Fordism may be more effective.

The Changing Focus of Vehicle Emissions Regulation

More than 1.4 billion cars and trucks travel the earth's roads today, and the size of that fleet is growing quickly. Transportation is responsible for 29 percent of GHG emissions in the United States (and nearly 50 percent in California) and about 21 per cent worldwide.[1] China surpassed the United States as the largest national contributor of GHG emissions around 2010, and although the burning of coal for heating and electrical generation has comprised the largest share of China's CO_2 emissions, the share from cars and trucks is now growing more quickly. Since 2000, China's promotion of the auto industry and of roadbuilding as a central part of its industrial strategy has made it the world's largest manufacturer of new cars and the biggest market for new car sales. If China reaches the same rate of ownership as the United States, that nation alone will have 1 billion cars and trucks. And as I have explained previously, the Chinese government's recent regulation of automobile emissions follows models established in Europe and North America, but also applies behaviorist methods that depart from consumerist ideology and evoke the legacy of Fordism.[2]

As concern about global climate change has increased since the 1990s, the major GHGs, CO_2 and methane, are now considered among the most dangerous forms of air pollution. At the start of the previous three-decade period, however, noxious air pollution from vehicles and factories and its impact on public health was the major concern about air pollution. The shift reflects not only research on global climate change, but also the success of automobile emission controls beginning in the 1970s. Emissions from gasoline and diesel-fueled engines of cars and trucks is a toxic mix of volatile unburned hydrocarbons, particulates, and carbon monoxide, as well as photochemical oxidants, which include oxides of nitrogen, sulfur dioxide, and the ozone that is created by the interaction of these emissions with bright sunlight. CO_2, which has been a major focus of European and American fuel efficiency regulations since around 2000, is not a direct threat to humans who breath it; it is the inevitable result of burning any hydrocarbon fuel, whether wood, coal, or petroleum products. Regulating transport as a source of GHG emissions (as US and European Union fuel economy standards attempt to do) aims to reduce a pollutant that had not previously been considered a threat to public health.

Regulating CO_2 in automobile emissions (using for instance the bonus/malus system in France) is only the latest twist in the evolving story of how cars have been understood as a source of pollution. In the early years of the car, before 1900, the new machines were regarded as an important advance for sanitation efforts in cities, because horseless carriages, as they were some-times called, held the potential to replace the millions of horses that pulled carriages and wagons. Each horse produced as much as 40 lb of dung every day, which on city streets combined with dust and dirt to create noxious smells and bacteria that posed a public health problem. But city dwellers who welcomed the removal of horses soon noticed how internal combustion engines emitted noxious "smoke," as it was often called, which added to the effluents of coal-fired heating and of industrial sources. The relationship between air pollution from industrial sources and from cars owned and driven by individual citizens will be a theme of the inquiries that follow.

To understand the significance of GHG emissions from transport compared with emissions from industry and from electric power generation, it is also important to acknowledge how, as the United States and Western Europe have become postindustrial economies, more and more of the factories making consumer goods (including cars and trucks) have shifted to China and South Asia. Some of the rich nations of North America and Europe have reported level or slightly declining GHG emissions, largely as a result of this trend, and of the increase in solar and wind-generated electricity. However, another consequence of the globalization of manufacturing since the 1990s is that shipping and transport accounts for a higher proportion of GHG emissions. Although container ships and cruise ships are notorious polluters, cars are the most ubiquitous and thus most polluting form of transport, and private vehicles' share of GHG emissions has risen both in developed and developing economies. And whereas electric vehicles are a cleaner, more efficient form of transportation than internal combustion engines, the enormous fleet of nearly a billion-and-a-half cars and trucks will only slowly be replaced by electric vehicles, and the sources of electricity to power such a huge new fleet have not yet been developed.

Fordism and Detroit

Cleaner emissions from cars and trucks can only be achieved, I argue, by addressing both the behavior of consumers, as owners, drivers, and passengers, and the labor and capital that produce the vehicles. To do so I turn to an influential economic concept from a century ago. "Fordism" in social science and industrial management derives from the methods that Henry Ford devised for building the Model T in 1908 at the Highland Park plant in Detroit, Michigan. The Ford Model T became the best-selling car in the world for nearly twenty years, a success that enabled Ford to expand its production to the River Rouge complex in Dearborn, which was the largest factory in the world when it opened in 1927. "The Rouge," as it was popu-larly called, continues to operate today, and now includes a new assembly plant for the F150 Lightning, the electric version of Ford's class-leading pick-up truck, which went on sale in 2022 and is a key test for the transition to electric vehicles in the United States. Fordism as a term in social theory was advanced by the Italian intellectual Antonio Gramsci in his *Prison Notebooks*, written in the 1930s, and became widely used by Marxist social scientists. Gramsci defined Fordism as

> the experiments conducted by Ford and to the economies made by his firm through direct management of transport and distribution of the product. These economies affected production costs and permitted higher wages and lower selling prices. ... [I]t was relatively easy to rationalise production and labour by a skilful combination of force (destruction of working-class trade unionism on a territorial basis) and persuasion (high wages, various social benefits, extremely subtle ideological and political propaganda) and thus succeed in making the whole life of the nation revolve around production.[3]

Fordism, whether a force for good or for evil, is generally judged by its effect on workers, not on the natural environment. But the story of the colossal pollution generated at Ford factories such as the River Rouge complex can also be understood through Fordism. In 1914 Henry Ford burnished his popular fame by announcing the Five Dollar Day for his factory workers. This was a higher wage than other Detroit factories were paying, and it paid off for Ford in reduced turnover, absenteeism, and training costs, as well as facilitating a regime of three eight-hour shifts in factories that could operate around the clock.[4] Henry Ford's assertion that his employees should be able to afford to buy the cars they made was not simply a recruitment slogan, nor a statement of his egalitarian ideals; it was a connection between the identity of a laborer and a consumer. Ford and his workers knew intimately how the Model T was made, and so they also knew how to drive and fix the car, and they knew the pollution produced in its manufacture. When workers work at the same site and live in the same city as managers, both suffer from the same pollution. Henry Ford built his stone mansion, Fair Lane, on the banks of the Rouge River beginning in 1914, and Ford employees who lived in Dearborn, adjacent to the River Rouge factory, later saw how toxic dust and ash covered their cars and pitted the paint on both cars and houses: "Soot rained down on the neighborhood [the South End of Dearborn] Hanging laundry out to dry was impossible, and people would not open their storm windows in the summer without coal soot piling up on their windowsills."[5]

Water pollution was just as severe. In the 1930s the Rouge plant poured 3,300 lb of cyanide into the river every day. "One Michigan water quality official drew a bucket of water from the Rouge, only to find that within ninety minutes the acid had eaten away the bucket's bottom."[6] Shared occupational, recreational, and health interests between workers, management, and political leaders in the Detroit area gave early antipollution activists some traction, an early form of what cultural anthropologists today call civic technoscience.[7] As historian Tom McCarthy wrote in his book on the history of the US auto industry and the environment:

> Hunters and fishermen worried companies like Ford a good deal. They were customers, and in Michigan, they were numerous. They also showed a willingness to embarrass the automakers publicly. When they complained in dealer showrooms about industrial pollution from the Rouge, the dealers told Ford it hurt sales. Many sportsmen worked for Ford, too, in both hourly positions and management ... Protesters and the plant engineers often shared this common identity and basis for respect, one reason why water pollution got more attention.[8]

These subtle protests made some progress toward reducing pollution in the Detroit area in the 1930s to 1950s, through tacit, nonenforceable agreements between men who worked in industry and in local government. Nonetheless, air pollution worsened because car sales continued to grow in the late 1950s and into the 1960s as many families owned more than one vehicle. Antipollution activism in the United States since the toxic waste scandal at Love Canal (in Niagara Falls, New York) in the 1980s, has often been led by women, mothers, and communities of color who are not employees of the institutions emitting the pollutants. This narrative continues to influence the environmental justice movement of today. The activists in Dearborn area in the 1960s, however, rose from among the workers at the River Rouge plant and nearby suppliers, a mix of European and Arab-American immigrant families. Vincent Bruno worked at the River Rouge foundry and was president of the South Dearborn Community Council. He and clergymen Paul J. Hardwick and Father Albert Lombardi had the idea to build a parade float depicting the River Rouge plant and its eight smokestacks. At the 1967 Memorial Day Parade in Dearborn the float was animated with children wearing gas masks and hanging laundry on clotheslines. The van towing the float bore a sign reading "Greeting. From the Valley of Death."[9] As the 40,000 spectators of the parade began to see it, local officials ordered police to force the float out of the parade before it reached the review stand where dignitaries sat. This spectacle got the attention of the media, and of Ford's management and local political leaders, and helped lead to the formation of a Wayne County air pollution control office, to coordinate pollution regulation in the Detroit area.

An Interlude on Automotive Fuels and Engine Pollution

To be precise, it is not car and truck engines that pollute, but the petroleum fuels those engines burn. And the technologies and corporate interests of petroleum refining have long been closely intertwined with automobile manufacturing. In the late nineteenth century the early petroleum industry produced kerosene for use in lamps, and dumped gasoline, then a useless by-product, into nearby streams. In the early 1900s gasoline pulled ahead of battery electricity, alcohol, and diesel as the most widely used vehicle fuel, in part because new refining techniques increased the yield of gasoline from crude oil, thereby keeping retail prices low. During World War I Allied powers depended on gasoline to power war machinery, and they conspired to secure control of Middle Eastern states with huge crude oil reserves. Meanwhile, General Motors (GM) researchers experimented with various fuel additives to try to improve combustion in the engines they designed, and in 1921 found that a small amount of tetraethyl lead in gasoline allowed for engines with higher compression ratios to run smoothly and produce more horsepower. GM and Standard Oil, the monopoly headed by John D. Rockefeller, created the jointly owned Ethyl Gasoline Corporation in 1924.[10] This corporate collusion encouraged GM to promote leaded gasoline for its cars for almost fifty years, even as medical research came to show the enormous damage of lead poisoning on human health, especially children's brain development.

While oil refining and auto manufacturing can be held responsible for the suffering of millions owing to lead poisoning, the industry also had its heroes. Eugène Houdry was an engineer whose career further illustrates the interdependence of the automobile and petroleum refining industries. Houdry studied engineering at the Institut des Arts et Metiers, then the most prestigious faculty in Paris, and became a decorated veteran of World War I, where he repaired tanks in an artillery unit. In the 1920s he formed a company to develop improved fuels for racing cars (such as the Bugatti he raced himself), and then in the 1930s he emigrated to the United States and continued his research at refineries in Pennsylvania and New Jersey. By the 1940s the high-octane aviation fuels his company invented were important for the Allies' war effort. After World War II, Houdry became concerned about the health effects of air pollution, and in 1956 received a US patent for a catalytic converter, a device installed in exhaust pipes that reduced the unburned hydrocarbons and sulfur in automobile emissions. Lead in gasoline clogged the catalysts in the devices, however, rendering them ineffective after less than 10,000 miles of use. Houdry knew that lead additives were no longer necessary to create fuel of an octane rating high enough for the engines then used in cars, but the US automakers, of which GM was the largest, continued to endorse leaded fuel and to resist implementation of catalytic converters.

Consumerism and Los Angeles

Detroit is known as "Motor City," a nickname that was adapted for the music genre Motown in the 1960s, and endured even as auto assembly and parts factories spread throughout the Midwest, and more recently the Southeast, where state laws make it difficult for unions to organize factory workers. By contrast, Americans recognize Southern California as the birthplace of popular artistic expressions of the automobile such as the hot rod and lowrider, and as the region most dependent upon cars and expressways. Imported Japanese cars began to gain market share in the United States in the 1970s, particularly in California, and the Los Angeles area developed design studios and training schools that support both Japanese and US manufacturers. The United States has evolved into a regionalized economy wherein the West and Northeast cities are centers for finance, computing, and design jobs, while the Midwest and South are known for heavy industry and agriculture. California's state government has not been strongly influenced by auto industry workers or lobbyists, the way Michigan and other Midwest states have been, and so California has played a key role in the history of auto emissions politics and regulation.

During World War II, Los Angeles faced its first air pollution emergencies: "a dirty grey blanket flung across the city, a dense eye-stinging layer of smog dimmed the sun here yesterday."[11] The cause of the smog was both car exhaust and the many local aerospace factories supplying weapons for the war in the Pacific. Rather than a product local people used for themselves, like the Model T, these factories made war machinery used to attack people on the other side of the planet. In an ironic reflection of the environmental impacts of war, many Angelinos imagined that a toxic air episode in July 1943 must have been caused by a sudden Japanese attack on the city. Chronic air pollution in Los Angeles in the 1940s and 1950s damaged public images of the region, which advertised itself to tourists and to migrants seeking to move to a sunny climate. As the Detroit three, GM, Ford, and Chrysler, consolidated their monopsony over the auto industry in the United States, Southern Californians were slow to accept the truth that it was their own cars polluting their air.

The geography of the Los Angeles region makes it particularly susceptible to severe air pollution. The metropolis sits in a basin surrounded by mountains on the north and east sides and by the Pacific Ocean on the southwest. Cool ocean breezes often clear the air near the beaches, but the mountains and the hot dry deserts beyond them create high pressure that prevents these winds from moving further inland and dissipating the city's pollution. The result is a temperature inversion wherein dense cool air from the ocean creates a layer above the warmer air of the city and traps its auto exhaust and other polluted effluents. The inversions can persist for weeks. Similar patterns now afflict other urbanized mountain valleys in the American West, such as Salt Lake City, and California's own central valley, but whereas inversions in those places occur mainly in winter, in the South Coast Air Basin (as California pollution regulators referred to the Los Angeles region) they can occur at any time of year. In the early 1950s these atmospheric phenomena were not yet understood, nor was the chemistry that created smog.

To resolve these questions, Arie Haagen-Smit, a chemist at the California Institute of Technology in Pasadena, a suburb of Los Angeles, did a series of experiments to determine the constituents of smog. The word was coined as a portmanteau of "smoke" and "fog," but in truth the acrid brownish gas consists of ozone, nitrogen oxides, and unburned hydrocarbons, catalyzed by sunlight in the air.[12] Angelinos believed clear air and sun defined their city, and they did not at first accept that their own cars and sunshine were the primary cause of the smog that afflicted them. While the aerospace and defense industries were major employers in Southern California, the automobile manufacturers were not. California was quicker than Michigan to begin regulating point-source air pollution emitted by factories and refineries. In the words of legal scholar Eli Chernow:

> Los Angeles County district had pioneered the field of air pollution control and had forced a number of regulations upon resisting interests. It had forced residents of Los Angeles to abandon backyard incinerators as a common method of rubbish disposal. It forced the oil refineries in the area to spend $150,000,000 on emission control equipment. Perhaps most importantly, it forced large scale fuel users, principally the public utilities, to burn low-emission fuels. As a result, Los Angeles, unlike many other American cities, did not have a serious sulfur dioxide pollution problem arising from the burning of coal or gas with high sulfur content.[13]

In the late 1960s urban California's most serious air pollutants were carbon monoxide, photochemical oxidants, nitrogen dioxide, and hydrocarbons, all of which were prominent in car exhaust. Factories with smokestacks are easy to locate and popular targets for regulation, unlike the dispersed pollution of automobiles, to which every driver is a contributor. Haagen-Smit held the unenviable task of persuading the state that clearing the air could only be done by drastically reducing the pollution coming out of automobile tailpipes. To accomplish this feat the catalytic converter became the most effective tool. A few months before Eugene Houdry's death in 1962, his company announced an improved catalytic converter that could greatly reduce toxic emissions. After learning of Houdry's breakthrough, in 1964 the Motor Vehicle Air Pollution Control Board (MVAPC) set out to fulfill the mandate set for the board by California law, which was to approve at least two devices that would reduce engine emissions. The board certified four, of which "None of the devices had been developed by the automobile industry, but by outsiders."[14] At that time the catalytic converter was just one concept among several being explored, including Wankel rotary engines to replace pistons and compressed natural gas to replace gasoline.[15] The action of the board was an early instance of what lawyer John Bonine later called "technology-forcing regulation," whereby regulators help break the resistance of profit-driven industrialists to research and implement new products or methods needed to protect consumers and the environment.[16] Haagen-Smit and others on the board saw their action as a way to call GM's bluff, or call out its intransigence. GM still owned half of the Ethyl Gasoline Corporation, and did not want to install catalytic converters, nor to sell cars that could only operate on unleaded gasoline. Fortunately, in the late 1960s GM executive Ed Cole saw the potential of the catalytic converter, and he confronted the oil industry to demand that lead be eliminated as a gasoline additive. Cole told the convention of the Society of Automotive Engineers in January 1970 that "It is my opinion that the gasoline internal combustion engine can be made essentially pollution free," which turned out to be only a modest exaggeration.[17]

Let's step back and consider some lessons of the catalytic converter. In 1964 both consumers and regulators in California wanted a device that would allow the same cars and same fuels to continue to be produced, sold, and driven, but to emit fewer pollutants. "Faced with changing their behavior—by driving less—or installing emission control technology on their cars to deal with the chemical causes of smog, Southern Californians turned their eyes toward Detroit."[18] Los Angeles germinated a consumerist movement that took aim at Fordist Detroit. The catalytic converter (together with the removal of lead additives and later changes in petroleum refining such as low-sulfur fuels) made such an engineering solution possible, and California regulators made it a reality. Automakers at first resisted it out of narrow-minded intransigence, but after their intransigence was exposed, smart politicians no longer believed the automakers when they claimed no engineering solution was available to meet a regulatory demand (such as air bags to improve crash safety or hybrid drivetrains to improve fuel efficiency). When consumer advocate Ralph Nader learned of the California MVAPC board's announcement, he got in touch with board member S. Smith Griswold and Los Angeles County supervisor Kenneth Hahn to discuss filing suits against the Detroit automakers. Nader planned to argue in court that the automakers were conspiring to delay implementation of emissions control devices such as the catalytic converter. Nader saw a precedent for this behavior because California regulators had in 1961 mandated an earlier engine innovation, the positive crankcase ventilation (PCV) valve, on all new cars sold in California. The PCV valve reduced unburned fuel emissions by rerouting part of the exhaust stream back into the combustion chambers. Detroit automakers grudgingly installed them on all vehicles, whether sold in California or elsewhere. Thus Haagen-Smit, Nader, and others sensed the industry was repeating its stalling tactics with the more sophisticated, costly, and effective catalytic converters.[19] Unlike the story of Ford's River Rouge factory and the ideology of Fordism, Nader's movement presented the interests of consumers as opposed to, not aligned with, the interests of corporations, their employees, and stockholders. It also helped make California the proving ground for automobile emissions regulation, from which new concepts would spread to the United States as a whole, and to other nations.

Ralph Nader soon became famous thanks to his bestselling book *Unsafe at Any Speed*, published in 1965. Most of the book concerns safety defects and poor designs of cars made by the Detroit three, but one of the eight chapters is about pollution and emissions, drawing upon Nader's collaboration with the Los Angeles researchers and activists. In the book the car becomes an agent mediating between the faceless corporation that manufactured and marketed it and the consumer who has placed faith and money in it. The first chapter and best-known part of *Unsafe at Any Speed* is an exposé of the 1960–1964 Chevrolet Corvair, a rear-engine, air-cooled mid-size model that GM designed as a response to the growing popularity of the Volkswagen Beetle. The rear suspension of the Corvair used a crude swing-axle design that allowed the camber of the wheels to change with the road surface, and many crashes resulted when a rear tire "tucked under" and caused the car to roll over during a turn, even at moderate speeds. Nader described the process as if the car were a gothic monster: "In ways wholly unique, the Corvair can become an aggressive, single minded machine. ... Rim scrapings or gouge marks on the road have become the macabre trademark of Corvairs going unexpectedly out of control."[20] GM's response to internal critics of the design was not to install a simple anti-roll bar (such as were used on many other car suspensions) but instead to advise buyers to inflate the front tires to a lower pressure than the rear tires, a directive unique to this model and inconsistent with the recommendations of tire manufacturers. In this melodrama, consumers are passive creatures of habit who cannot be expected to learn new habits or skills, whether to protect themselves or to improve their society's health and well-being. Instead, consumers simply purchased and used products, such as the Corvair, that could be wily, deceptive, unpredictable characters who did not care for their owners as strongly as the owners cared for them. Nader and other muckraking journalists wrote in an anti-romantic mode that relished in exposing the nefarious deeds of capitalists who seduced and betrayed consumers. The ethical and legal arguments Nader made in the book helped to bring about key environmental protection laws enacted by the US Congress over the following decade. However, his appeal was predicated on an unrealistic premise: Consumers' behavior need not change, because it was the products, and the companies making and selling them, that were at fault.

The weakness of Nader's consumer protection movement was that it reified defective or dangerous products, rather than considering consumers and products as interactive or conspiring agents in social/environmental problems. To build on this point we should consider toxic emissions as just one of many externalities of automobiles. Externalities, a term used in economics, describes how a market actor such as a consumer benefits from a product or activity, while costs, burdens, or dangers—for which that actor is not held liable—are imposed on others.[21] A car driver and passengers benefit from personal transport, and all others external to that car suffer from the GHG emissions, the traffic congestion, the lost space consumed by highways and parking lots, and the increased risk of collisions and injuries. Consumer activism on behalf of car owners generally does little to alleviate these externalities. Making a car safer by installing airbags, antilock brakes, and the myriad warning chimes and self-driving features of new cars in the 2020s, may reduce risks for that car's occupants, but often does little to protect pedestrians and other noncar users on the road. Indeed, car crash fatalities in the United States have risen steadily since 2015, with deaths among pedestrians, bicyclists, and motorcyclists growing much faster than for car drivers, even as some European countries have seen overall improvements in road safety.[22]

The catalytic converter was a greater benefit to the public than other features safety and emissions regulations have brought to cars since such regulations began, with California's leadership, in the early 1960s. Catalytic converters were installed on more than 85 percent of 1975 model year cars sold, with the result of reductions in unburned hydrocarbons by 83 percent, carbon monoxide of 83 percent, and nitrogen oxides by 11 percent.[23] In the next five years, automotive engineers at the Detroit three collaborated with Volvo to develop a three-way catalytic converter that also reduced nitrogen oxides. The success of the catalytic converter lay not only its chemical details, but also in the structure of air pollution and the car industry. Although cars are dispersed, unlike factories emitting air pollution, they are nearly all manufactured by a few huge corporations, whose grudging agreement to cooperate with regulators quickly changed the entire industry. Reduced pollution is a benefit to all, for there is no rivalry or exclusivity in the consumption of air, to use two more economic terms. One person breathing does not reduce the air available to others, and no one is obliged to pay for what they breathe. Everybody in a regional airshed benefits from such emission controls, whether or not they come into contact with any particular car. And whereas many recent safety features have been introduced as extra-cost options, the catalytic converter was mandated for nearly all new cars beginning in the 1975 model year. Recent cultural anthropology research has developed "Chemo-ethnography," defined as "studying how initiatives to monitor water quality or to bioprospect for pharmaceuticals have displaced other pressing community concerns and limited conversations to the priorities of scientists, policymakers, and corporations."[24] The story of the catalytic converter constitutes an early chemo-ethnography of an invention that changed the course of automobility and politics in profound ways that few today are aware of.

Although the catalytic converter greatly reduced toxic air pollution, a truly comprehensive environmental analysis must consider the resources required to make it, and the essential catalysts in the converter are platinum and palladium, two rare metals of which more than half the global production is now used for that purpose. In the early 1970s GM worked with South African suppliers of platinum who opened a new mine to exploit a geological feature known as the Merensky Reef, with mining companies hiring 10,000 miners who worked in apartheid-era conditions, in jobs much more dangerous and poorly paid than those in car factories. These impacts were out of sight and out of mind for American drivers, but should not be ignored in considering the environmental impacts of cars and engine emissions.

Ralph Nader's influence and skills as an activist went beyond the consumerist ideology initiated with his work for the catalytic converter and against the

Corvair. Nader sought to extend the concept of price fixing (whereby corporate interests collude rather than competing on price) to "product fixing," whereby the four major automakers colluded to hold back safety or emission-control features from their cars. The concept caught on in legal circles, and the US Justice Department antitrust division under President Lyndon Johnson in 1966 began an investigation against the Automobile Manufacturers Association (AMA), and empaneled a grand jury to collect evidence. In the eyes of Nader and federal lawyers who embraced his theory, a collaborative effort of the automakers to develop emission control technologies, known as the Inter-Industry Emission Control Program, was in truth a plan to collude and delay the adoption of devices such as the catalytic converter. In 1969 Attorney General John Mitchell, appointed by President Nixon, reached a consent decree and an out of court settlement; the AMA promised not to withhold emissions control technology henceforth, and all the evidence the grand jury had collected was sealed. This was the end of United States vs. Automobile Manufacturers Association case in the US District Court in California.

Many saw the outcome as a cover-up. The automakers were able to hide evidence of their wrongdoing and avoid a public trial. A research group formed by Ralph Nader, the Center for Study of Responsive Law, published a lengthy report in 1970 entitled *Vanishing Air*, which stoked suspicion that the consent decree was indeed a cover-up of criminal collusion.[25] On top of the suspicions aroused by the investigation, smog that choked Southern California, as well as New York, Philadelphia, and Washington, DC, in the summer of 1970 provoked a storm of protest from members of Congress. Public alarm about air pollution was at a high point, and political habits shifted. President Richard Nixon, a lifelong pro-business Republican, oversaw the creation of the Environmental Protection Agency (EPA) in December of that year, and encouraged a Congress led by liberal Democrats to write and pass some of the strongest environmental protection laws the United States has ever seen. There is not space here to explain the full sequence of legislation on air quality since the early 1960s, but the most important for our story are the amendments to the Clean Air Act, written largely by Senator Edmund Muskie of Maine, which passed the House 375–1 and the Senate 73–0 in the summer of 1970. The new law set a strict timeline for every city and state to reduce the number of days per year that ambient air exceeded limits for each of a list of major pollutants, and draft detailed plans for how to reduce pollution. The EPA and its administrator, William Ruckelshaus, were granted extraordinary powers to enforce the law. As legal scholar Eli Chernow explains,

> The EPA Administrator was to review all implementation plans, rejecting those plans or parts thereof which failed to meet the Act's requirements. In addition, the Administrator had the explicit affirmative duty of promulgating substitute plans, or parts of plans, when the plans submitted were found to be insufficient.[26]

In Southern California over the next two years, the strict provisions of the law came up against the realities of automobility, in a confrontation that is not widely remembered fifty years later but has important lessons for efforts to limit GHG emissions today. The Southern California Air Resources Board, still chaired by Arie Haagen-Smit, was tasked with drafting an implementation plan by January 1972 that would enable the region to meet air quality standards. This was a daunting task given that Los Angeles County had exceeded the limits for photochemical oxidants on 241 days in 1970. Catalytic converters were still in development, and so notwithstanding Ed Cole's claims, there was no assurance that an engineering fix to cars would enable Americans to continue to drive inefficient vehicles everywhere they went and still breathe healthy air. On the other hand, Los Angeles politicians such as Kenneth Hahn did not wish to implement even modest efforts to reduce driving, such as new taxes on gasoline and parking, which they knew would arouse a fierce backlash. The EPA itself commissioned a study in Los Angeles and found that "even a free rapid transit system that got all Angelinos where they wanted to go as fast as the automobile would reduce automobile VMT [vehicle miles traveled] by only 10 percent."[27] Hahn and other California leaders believed the confrontation could be productive, however, if it forced the automakers to deliver on their promises. EPA chief William Ruckelshaus knew the law demanded that he reject the plan the CARB had submitted, but he also expected Congress to recognize how ambitious, or rather impossible, the 1970 Clean Air Act amendments were, and to quickly revise the law. From January to November 1973 the EPA, acting through Ruckelshaus, presented three alternative plans for the South Coast Air Basin. The first and most draconian was based on an estimate that the air would be clear of all the pollutants if gasoline consumption were reduced by 80 percent, through a rationing program. Public hearings concerning this plan enabled Los Angelinos to vent their outrage, and the EPA to then adopt a conciliatory posture. A second and a third alternative withdrew the threat of gasoline rationing and offered additional detail on the transportation plans, such as dedicated bus and carpool lanes on freeways and surface streets, and steep taxes on parking to fund improved bus service. Ruckelshaus called it "a national test case in the willingness of citizens to alter lifestyles to meet air quality standards."[28]

Updating Fordism and Reforming Consumerism

The lessons of the twentieth century's battles against air pollution from automobiles and the factories that made them are relevant for the challenge the world faces now from GHG emissions. CO_2 and methane are in some respects a new type of pollution, emitted from both natural and anthropogenic sources, and subject to complex feedback loops involving climate, vegetation, and ocean currents. But with regard to the 1.4 billion automobiles and their enormous GHG emissions, the same problems of externalities and consumer behavior arise now as when the EPA and the Clean Air Act confronted Southern California smog in 1973.

The story of how American activists, politicians, regulators, and consumers confronted the smog caused by auto emissions in Southern California might be seen as an anticipation, on a small scale, of the political efforts to address GHGs and climate change at the United Nations Climate Change Conferences in Rio de Janeiro, Paris, and Glasgow. There was a similar dialectic between a high-minded resolve to protect human lives and health, and a bitter backlash on the practical level as consumers and politicians, who had hoped or tried to shift the burden onto others, faced the prospect of enormous and immediate changes to their daily habits. The analogy with the current politics of GHG emissions falls short, however, when one considers the catalytic converter and related methods to reduce emissions from automobiles. These engineering solutions have succeeded beyond the expectations of the activists, scientists, politicians, regulators, and industry executives from the 1960–1975 period. The toxic pollutants in auto exhaust have declined more than 90 percent from the levels in the pre-emission control era, in the countries where emissions regulations are enforced. On this score the comparison with GHG emissions today is untenable. CO_2 emissions are an inevitable product of burning fossil fuels or even organic material such as wood. Methane is a by-product not only of fossil fuel extraction, but also of agriculture and of the melting of permafrost in Arctic regions. The equivalent of a catalytic converter to control CO_2 emissions would be a device attached to exhaust pipes that filtered out and transformed carbon into a solid before it could combine with oxygen. The carbon could then be buried to prevent oxidation. Even if this were chemically possible, on long trips the bulk and weight of the carbon payload could displace passengers and freight in cars and trucks.

Notwithstanding these fundamental differences between the story of auto emissions control for smog, and GHG emissions from the same transport sources, some lessons can be gained from the thought experiment.

First, the current method of imposing "technology-forcing regulations" upon cars is much the same as was devised in the 1960s. In the European Union CO_2 emissions from new cars and trucks are measured and limited through a corporate average fuel economy scheme similar to that first used in the United States for fuel economy in the 1970s. Hybrid and electric vehicles are incentivized through the credits and penalties to manufacturers building such cars, and often tax credits to consumers who buy them. The behavioral or urban planning methods that were so harshly rejected in Los Angeles, such as rationing of gasoline, taxes on parking fees and parking lots, or even incentives to use public transit, have rarely succeeded anywhere in the world, even in authoritarian regimes such as China.

Second, the technocratic expectations invested in the electric car today are in some ways equivalent to the catalytic converter in the early 1970s. Electric cars are seen as a magical product that will solve the GHG pollution problem without a need for consumers to drive less, use public transit, or pay punitive taxes on fuel. Government regulations and local pledges and plans declare that only electric vehicles will be permitted for sale within ten to twenty-five years. The evident success of emissions controls at improving air quality in North America, Europe, and now in China provide conceptual and political precedent in support of these plans. Electric cars have indeed improved in range and declined in price since 2010. But capacity for building electric cars is limited, as is consumer appetite for them. Moreover, to quickly reduce emissions requires measures that affect the entire fleet of vehicles, not just new ones, and policies developed recently do nothing to change older vehicles. When California issued its implementation plan in 1972 it made optimistic predictions for policies such as retrofitting existing vehicles to run on compressed natural gas, but consumerist ideology prevents regulators from modifying existing cars and trucks owned by individuals.

Third, the limits and less visible effects of the planned transition to electric cars are rarely recognized, and these invoke the Fordism of the twentieth century. Henry Ford was famous for his insistence on vertical integration of the supply chain for the Model T, buying up forest lands around the Great Lakes to supply the hardwoods from which its wheels and bodies were built, as well as mines, cargo ships, and even rubber plantations to supply metals and tires. Elon Musk's Tesla corporation, as well as GM, are now looking to develop lithium mines in Nevada and Oregon, because that metal is as critical for batteries as platinum and palladium are for catalytic converters. The resources required for new products such as electric cars detract from their environmentalist appeal, because new mines will likely create new sources of chemical contamination of air and ground water.

In spite of these demerits, consumerism remains ascendant in the US economy, and also entrenched in economic theory owing to the marginalist paradigm that defines value as consumer desire, rather than measured by labor and surplus value as Karl Marx did.[29] Fordism in some senses is long gone, because consumers—car buyers and drivers—know little of the workers and factories where their cars are assembled or the countless components from which they are made. Consumers no longer see themselves as sharing the interests of the companies that design the cars they drive, nor the workers who build them, and unlike owners of the Model T a century ago, they don't even feel responsible for understanding how their cars work.[30] I propose therefore a neo-Fordist paradigm for pollution control by which drivers regard themselves not only as consumers of cars and trucks, but also as producers of pollution. Because globalized sweatshop production has weakened labor unions and destroyed the Fordist principle that workers can buy what they make, I emphasize it is not truly factories that pollute; it is consumers themselves who produce waste, notably in the form of GHG. The concept of "apprehension" proposed by Rob Nixon captures the combined sense of anxiety and culpability of this neo-Fordist approach to consumer behavior.[31] As global warming becomes a focus of widespread fear, consumers should apprehend their carbon emissions and liability for GHGs, and recognize the externalities and indirect wastes of their consumption of all kinds of goods (offshore industrial production and waste disposal, mining of lithium and rare earth metals, and the production of agricultural fuels such as palm oil, wood chips, and ethanol).

Neo-Fordism could encourage consumers to recognize their role as producers of waste and emissions, and to embrace incremental, broad-based, low-cost efforts to reduce vehicle miles traveled and encourage nonmotorized or zero-emission transport and walkable, green designs for cities and residential communities. As history of engineering scholar Peter Norton has shown, high-tech, futuristic engineered solutions seduce consumers and regulators alike, but the track record for electric and hydrogen-powered cars was for decades a story of failure and delay.[32] Instead, we need to work toward what Norton terms "transportation sufficiency," through more careful use of the cars we have now, because we cannot depend on salvation from an engineering solution such as occurred with the catalytic converter fifty years ago.

Notes

1   *Global Warming on the Road: The Climate Impact of America's Automobiles* (Environmental Defense Fund, 2006), iv.

2   Gordon M. Sayre, "'Carbolization': Cars, Carbon Emissions, and the Global Discipline of Automobility," in *Transportation and the Culture of Climate Change: Accelerating Ride to Global Crisis*, ed. Tatiana Prorokova-Konrad (Morgantown: West Virginia University Press, 2020), 83–102.

3   Antonio Gramsci, *Selections from the Prison Notebooks* (New York: International, 1971), 570–71.

4   See Robert Lacey, *Ford: The Men and the Machine* (Boston: Little, Brown, 1986), 117–29.

5   Tom McCarthy, *Auto Mania: Cars, Consumers, and the Environment* (New Haven, CT: Yale University Press, 2008), 161.

6   McCarthy, *Auto Mania*, 110.

7   See, for example, Sarah Ann Wiley, Kurt Jalbert, Shannon Dosemagen, and Mark Ratto, "Institutions for Civic Technoscience: How Critical Making is Transforming Environmental Research," *The Information Society* 30, no. 2 (2014): 116–26. https://doi.org/10.1080/01972243.2014.875783

8   McCarthy, *Auto Mania*, 113–14.

9   McCarthy, *Auto Mania*, 162.

10 John C. Esposito, *Vanishing Air: the Ralph Nader Study Group Report on Air Pollution* (New York: Grossman, 1970), 65–68.

11 *Los Angeles Times*, 1946, quoted in Esposito, *Vanishing Air*, 36.

12 McCarthy, *Auto Mania*, 118–23; J. Robert Mondt, *Cleaner Cars: The History and Technology of Emission Control Since the 1960s* (Warrendale, PA: Society of Automotive Engineers, 2000), 4–13. https://doi.org/10.4271/R-226

13 Eli Chernow, "Implementing the Clean Air Act in Los Angeles: The Duty to Achieve the Impossible," *Ecology Law Quarterly* 4 (1973–1974): 541.

14 Esposito, *Vanishing Air*, 40.

15 For a sense of how the proposed devices were understood at the time, see Wesley Marx, "Los Angeles and its Mistress Machine," *Bulletin of the Atomic Scientists* (April 1973): 4–6, 44–48. https://doi.org/10.1080/00963402.1973.11455465

16 John Bonine, "The Evolution of 'Technology-Forcing' in the Clean Air Act," *Environment Reporter Monograph* (1975).

17 Jerry M. Flint, "GM Sees Autos Fume-Free by '80," *New York Times*, January 15, 1970, accessed July 27, 2022. https://www.nytimes.com/1970/01/15/archives/gm-sees-autos-fumefree-by-80-company-head-feels-us-will-set.html?smid=url-share

18 McCarthy, *Auto Mania*, 122.

19 See Esposito, et al., *Vanishing Air*, 40–47.

20 Nader, *Unsafe at Any Speed: The Designed-In Dangers of the American Automobile* (New York: Grossman, 1965), 33. https://doi.org/10.2307/251051

21 See Richard C. Porter, *Economics at the Wheel: The Costs of Cars and Drivers* (San Diego, CA: Academic Press, 1999).

22 Madeleine Ngo, "Transportation Dept. Outlines Plan to Address Rising Traffic Deaths," *New York Times*, July 27, 2022, https://www.nytimes.com/2022/01/27/us/politics/dot-traffic-deaths-plan.html?smid=url-share, accessed July 27, 2022; Emily Badger and Alycia Parlapiano, "The Exceptionally American Problem of Rising Roadway Deaths," *New York Times*, November 27, 2022, accessed February 4, 2023. https://www.nytimes.com/2022/11/27/upshot/road-deaths-pedestrians-cyclists.html?smid=url-share

23 McCarthy, *Auto Mania*, 189.

24 Nicholas Shapiro and Eben Kirksey, "Chemo-Ethnography: An Introduction," *Cultural Anthropology* 32, no. 4 (2017): 482. https://doi.org/10.14506/ca32.4.01

25 The abortive lawsuit against the auto industry was less effective than those brought against the US tobacco industry, settled in 1998, and against drug companies and pharmacies marketing synthetic opioids, settled in the early 2020s. In those cases, state Attorneys General were able to extract large payments from the product manufacturers, to compensate victims and provide for remediation.

26 Chernow, "Implementing the Clean Air Act in Los Angeles," 541.

27 McCarthy, *Auto Mania*, 205.

28 McCarthy, *Auto Mania*, 194.

29 Ronald L. Meek, "Marginalism and Marxism," reprint in *The Marginal Revolution in Economics: Interpretation and Evaluation*, ed. R.D. Collison Black, Alfred William Coats, and Craufurd D.W. Goodwin (Durham, NC: Duke University Press, 1973), 233–45.

30 See E.B. White, "Farewell, My Lovely," *New Yorker*, May 16, 1936.

31 Rob Nixon, *Slow Violence and the Environmentalism of the Poor* (Cambridge, MA: Harvard University Press, 2011), 15. https://doi.org/10.4159/harvard.9780674061194

32 Peter Norton, *Autonorama: The Illusory Promise of High-Tech Driving* (Washington, DC: Island Press, 2020).

2

Dirty Air: Literary Tropes
of the Canadian Nation

Christian Riegel

The history of active European interests in what is now Canada reaches back
to the early seventeenth century, but the notion of a sea-to-sea nation bound
by common interests only became a serious concern in the mid-nineteenth
century. From east to west, Canada spans 7,821 km,[1] and until technological
developments made faster travel possible, crossing this span of geography
was accomplished primarily through a series of waterways and horse tracks
and relied mainly on human power. The construction of a railroad in the late
nineteenth century allowed for the understanding of a series of territories
that could be combined to create a nation and ended the possibility of the
United States encroaching upon northern lands. Canada as a nation came
into existence in 1867, and though a cross-country railway was completed
in 1885,[2] it was only in 1965 that a roadway, the Trans-Canada Highway,
connecting the easternmost point of Canada with its western counterpart,
was completed.[3]

My interest in this chapter is to examine a series of poetic and fictional
texts written over the last fifty years that engage with the complex mix of
colonialism, technology development, and nation-building that are connected
to the construction of the Canadian nation. Being born as it is on land already
inhabited by Indigenous people comprising approximately fifty different First
Nations,[4] Canada exists in an ongoing state of tension over its creation
reflected in a complex series of social challenges including racism, immense
poverty, lack of access to social services, and continuing land-related conflicts
related to the status of Indigenous people in the nation. Key to the challenges
of the developing Canadian nation are advancements in technology, including
the development of the railway and the Trans-Canada Highway, but also the
industrialization of mining, which was enabled by the arrival of the railway.

Christian Riegel, "Dirty Air: Literary Tropes of the Canadian Nation" in: *Imagining Air: Cultural Axiology and the Politics of Invisibility*. University of Exeter Press (2023). © Christian Riegel. DOI: 10.47788/FRQU2721

And, of course, with industrialized processes, be they railways, roadways, or mining smelters, come considerations of polluted air. Nation-building is inextricable from environmental degradation, and much of that degradation can be captured by exploring links to dirty air. Furthermore, the process of nation-building allowed for the subjugation of Indigenous people, including through the spread of airborne diseases—intentionally and negligently—causing the near-eradication of Indigenous populations alongside cultural and social degradation. The effects of nation-building on Indigenous peoples in Canada are myriad in their negativity. The release of the Truth and Reconciliation Commission report in 2015 promoted nationwide recognition of the intensely negative effects of the residential school program that forcibly removed children from their parents for the purposes of assimilation, causing untold numbers to die of disease, neglect, and abuse.[5] Air is closely linked to technology development and the construction of the Canadian nation, which in turn led to many harms to Indigenous populations.

Dirty Air and the Canadian Nation

Nation-building in Canada is closely related to what can be extracted from the land or to what the land can be used for. Inherently, then, this leads to considerations of air—filthy, noxious, dangerous, unhealthy, air such as was produced in mining towns across the country. This sense of air as polluted serves metaphorically to indicate the state of the nation and its colonially rooted exploitation of Indigenous people alongside the environmental damage caused by the extraction of resources and the massive retasking of landscapes for agricultural uses (such as the grasslands of the western prairies, which have largely been destroyed owing to farming). My purpose is to examine how the metaphor—appearing in various guises—is deployed by a series of Canadian authors writing from various subject positions, including Indigenous Canadian.

Consider, for example, the Hudson Bay Mining and Smelter Company, which is fictionalized in the novel *The Red-Headed Woman with the Black Black Heart*, by Birk Sproxton,[6] a book to which I will return later in this chapter. The mine operated from 1927 to 2010 and was able to do so only because of the construction of a railway that connected it to the Canadian Pacific Railway (CPR), the railway that joined the Atlantic coast to the Pacific coast and allowed territorial unity in northern North America, leading to a nation being created. In historical accounts of the mine and of life in Flin Flon, which existed only to service the mine, living in a state of polluted air is normalized among the town's inhabitants, as one journalist remarks:

> Inside, the smelter relied on a process of heating and melting to extract copper from ore. The resultant pollution – toxic clouds of white smoke – poured out of the tips of two smoke stacks.
> Most Flin Flon residents accepted the smoke – and the occasional hacking it produced – as inherent to life in an industrial community.[7]

Indeed, the pollution created by the mining operation was understood by Flin Flon's inhabitants to be inherently unhealthy as "they worried about the many tonnes of lead, arsenic, mercury, and other dangerous pollutants in their air because of the smelter."[8] Further, environmental degradation went hand in hand with poor treatment of workers and poor living conditions in Flin Flon, leading to the 1934 mining strike[9] that forms the focus of Sproxton's novel. As the narrative of the Hudson Bay Mining and Smelter Company suggests, transportation technology and polluted air are closely aligned in the process of nation-building.

Colonialism, settlement, and exploitation of natural resources are intertwined in the history of Canada, as is evident in the case of the Flin Flon mine, and the metaphorical configuration of dirty air in literary texts can be seen to be grounded in the long history of industrial pollution that is associated with the nation. Pollution is fundamentally connected to colonialism, as Max Liboiron notes: "pollution is not a manifestation or side effect of colonialism but is rather an enactment of ongoing colonial relations to Land. ... [P]ollution is best understood as the violence of colonial land relations."[10] Marked by acts of conquest and genocide, colonialism provides "the foundation of colonial state-formation, settlement, and capitalist development on the other."[11]

The case of the Aamjiwnaang First Nation's reserve is illustrative. Located in Sarnia, Ontario, on the border with Michigan, the reserve has been defined as being located on "a toxic petrochemical complex. With over 60 refiners and chemical plants, 40% of Canada's chemical industry can be found here" and is "ranked as having the worst air-pollution in the country" by the World Health Organization.[12] Oil and chemical production dates from the 1850s,[13] and is defined as part of "the quest to colonize 'unruly', 'immoral', 'chaotic' or 'ungodly' spaces ... [to] justify an insatiable desire for advancement and perfection. The limits of progress and production in Canadian society are exposed in Aamjiwnaang," reinforcing the understanding that "toxic pollution" underpins nation-building.[14] Similarly, the long-standing and ongoing industrial development of the oil sands in northern Alberta has been well documented in terms of the human and ecological damage of pollution related to air and other sources. Inhabitants of the area report higher levels of respiratory illnesses, skin reactions, and cancer,[15] and "many other aspects of local inhabitants' lives are affected by industrial pollution, such as food security, [and] cultural practices and beliefs."[16] Lena Gross connects pollution of the oil sands to settler colonialism and the developing nation, noting that "[i]n the context of settler colonialism, dispossessions by pollution are signs of double domination, where land first gets appropriated and afterwards is made inaccessible to its original inhabitants to 'protect' them from the effect of the destruction to which it was subjected." Pollution should be seen as "an act of settler violence and a tool for dispossession,"[17] which is consistent with Liboiron's observations that nation-building and pollution are inextricable.

Polluted Air, Nation-Building, and Transportation Technology

Two literary deployments that illustrate some of the ways in which writers articulate a discussion of nation-building, technology, transportation, and polluted air serve as good entry points to such considerations. In her poem "Zone: <le Détroit>," Manitoba poet Di Brandt explores what it is like to live in the city of Windsor, Ontario, which sits near the mouth of the Detroit River directly across from the much larger city of Detroit, Michigan.[18] Both cities are deeply implicated in the auto industry, producing millions of carbon-burning vehicles over many decades. Brandt connects the state of the air to colonization of North America, and thus also to the development of Canada. As she writes:

> Breathing yellow air
> here, at the heart of the dream
> of the new world,
> the bones of old horses and dead Indians
> and lush virgin land, dripping with fruit
> and the promise of wheat,
> overlaid with glass and steel
> and the dream of speed:
> all these our bodies
> crushed to appease[19]

The processes of constructing a nation are linked to the unnaturalness of air that is "yellow" and the transformation of "virgin land" that was filled with a lush and providing nature but is instead made into an industrialized space that attends primarily to the needs of transportation—of bodies and goods. The new world is less *new* than it is refashioned through an ideology that understands progress primarily as destruction resulting in "crushed" bodies. The nation is thus "poisoned country," as Brandt writes.

Among the industrialization the poet finds in her sojourns in the cityscape are signs of a preexisting ecosystem:

> So there I am, sniffing around
> the railroad tracks
> in my usual quest for a bit of wildness,
> weeds, something untinkered with.[20]

The connection to the railroad and the juxtaposition of its industrial nature and the "wildness" that intrudes upon it resonates with considerations of the railway as nation-building. Nature has been tinkered with, as Brandt puts it, and though it is not entirely removed it is shifted to the margins by the necessity of industrial activity.

Likewise, in his 1997 book *Apparatus*,[21] Ontario poet Don McKay is interested in the relationship of transportation technology to a sense of nation in his three-poem suite "Three Eclogues." The concept of an eclogue is invoked ironically as the speaker-poet navigates a series of rural landscapes that are overlain with industrialized qualities. Each of the three pieces works through fundamental elements of nation-building via transportation, beginning with canoeing on the Raisin River in Ontario, reflecting early exploration and the fur trade, which relied entirely on waterways, moving then to a poem about the Trans-Canada Highway and its role in the movement of industrial and consumer goods, and finishing with an exploration of a railway scene.

In the first poem, "Sunday Morning, Raisin River," the speaker contrasts human-powered motion (which was also an Indigenous form of transportation) with the "outboards" that intrude upon the river and its banks reshaped by domesticated lawns. "To own, to mow," notes the speaker as a play upon the essence of colonially rooted uses of the land that emphasize private ownership. The boat motors are identified by their commercial names—the "Evinrudes and Mercs" indicate the brand names that fit with industrial production of consumer goods. The canoe allows access to what the poet describes as "the scruff of habitat," that place where the transformation of geography is natural rather than destructive and polluting: "Think how a reed or stick / can turn into a Great Blue Heron lifting off."[22]

The second in the series, "On Foot to the Bypass Esso Postal Outlet," most closely reflects upon my themes of the relationship of transportation to concepts of nation. The juxtaposition of human power, via the walking speaker and the bypass highway, reinforces the differences between the observing individual and the structures of nationhood. The post office, which is a federal institution, is combined with a private oil company and its fuel station, and it is located on the Trans-Canada where it "achieves four lanes" for maximum capacity. The trucks parked at the station represent large transportation companies and reflect the destructive elements of the carbon-based movement of goods as they are "always on the verge / of catching and devouring the future."[23] The processes are commercial, unnatural, and industrial. The trucks "have been everywhere / and boxed it, they've been hauling all the way / from depots in the late Cretaceous."[24] The mere human, along the side of the highway, "is dismissed"—irrelevant to the forces of development and nation. The detritus of these forces is evident in the ditches: "plastic allsorts, / chrome, / Styrofoam like pasteurized snow, cups from / Tim Hortons, one sneaker."[25] The objects reflect items that are transformed and/or contained, such as the snow and the cardboard coffee cup, and all are ultimately useless, like the single shoe. This is what motorized transportation brings to hitherto natural spaces. The speaker imagines speaking to a raven, the discussion ranging from the inane considerations of weather and hockey to "inconspicuous consumption" and "the/ importance / of the Trans-Canada as provider of dead meat."[26] Nationhood is bound to negative forces of transformation such as these.

The third poem in the series, "Abandoned Tracks," follows a similar pattern, exploring a railway and the poet-speaker's childhood memory of it.

The arrival of a train is marked by violence that punctuates the landscape and the people that inhabit it: "we'd feel it / in the air the way, I guess, a horse can sense an earthquake / coming." Unnatural, destructive, a "dreadful blundering," the train lugs "tanker cars full of polysyllabic stuff" as it powers through the landscape. It undertakes the work of nation-building, contributing to industrial development all the while displacing nature.[27] In the speaker's present, he notes that the railway has been abandoned. The tracks have been taken over by wild plants—"Now / Cow Vetch and Mustard get in the way / and hide the ties." What is present is a reminder of what was destroyed by the construction and operation of the railroad. As the speaker remarks in his description of the building of the tracks, "The spikes" are "hammered like cold bolts from the blue."[28] The absence of the railroad literally reinvigorates the landscape and also serves as metaphor for what is lost through the industrial development of the nation. The regrowth of vegetation reinforces the negative effects reflected in the metaphor of nation-building, the railway, and industrialization: "Everything the tracks / have had no use for's happening / between them."[29]

Railway, Nation, and Culture

How, then, is the creation of a national railway linked to the concerns that poets such as Di Brandt and Don McKay address? To begin to answer this question it is worth considering the forces that brought about transportation infrastructure. In the mid-nineteenth century, industrialists and others "promoted the construction of new transportation links as nation-building projects. These engineers, promoters, and politicians believed that enhanced mobility and communication could forge a new nation in northern North America."[30] Such considerations of nation-building persist into the contemporary moment, noting that "choices concerning mobility—the movement of people, things, and ideas—have shaped Canadians' perceptions of and material interactions with their country."[31] Globally, "the speed, distance, and regularity of corporeal movement began to increase on a scale unprecedented in human history, as a new energy regime took hold—with fossil fuels powering locomotives, steamships, and other modes of transportation that, in many places, supplanted older muscle- and wind-powered modes."[32] With the potential of fossil fuel driven transportation, it is not surprising that Canada's history is bound up in the broader global movement: "Given that the emergence of the Canadian nation-state had coincided with this transportation revolution, it must have seemed logical to many Canadians in the late nineteenth century to link improved mobility with national progress as though they went hand in hand."[33] This sense of national progress is bound up in narratives of the Canadian nation, eliding the environmental degradation and human suffering that accompanied it: "[T]ransportation and travel take the role of handmaidens in a romantic narrative of national ambition, economic development, scientific enlightenment, and material progress."[34]

In Canadian history this is often seen as "an epic struggle to penetrate the wilderness, capture resources, and consolidate the country through improved transportation [that] lives on in the popular imagination";[35] "Canadians continue to embrace mobility as symbolic of the desire to master time and topography."[36] Developing a railway that spanned the soon-to-be nation was imperative to the creation of a Canada, and this creation is predicated upon exploitation of natural resources and repurposing of land for agricultural purposes: "The Intercolonial Railway, a provision of the British North America Act that created the new Dominion of Canada, was completed in 1872 to provide access for central Canadian resources to the Maritime ports."[37]

The pervasiveness of this narrative of nation-building as progress is perhaps best articulated in a television series produced by the Canadian Broadcasting Corporation (CBC) in 1974 about the development of the CPR:

> In the opening sequence ... the pristine majesty of the Rocky Mountains and a lone Indian are confronted with the technological dynamo of a locomotive. This television image of a railroad as the "national dream" heroically spanning the wilderness to fashion a state reveals in a condensed narrative the manifold relations between technology and a Canada which can imagine.

Technology is defined as constituting Canada itself as a nation and is "a manifestation of Canada's ethos." The National Dream reinforces that the railroad was necessary to bring together the vastness of space that comprises the northern parts of North America:

> This CBC epic reminds us that Canada exists by virtue of technologies which bind space and that the railroad permitted a transcontinental economic and political state to emerge in history. Furthermore, the National Dream is an instance of the discourse of technology in Canada, of its rhetoric. The CPR is presented as the archetypal instance of Canada's technological constitution. More significantly, the CPR is offered as the product of political will. A nation and railroad were "dreamt" of by Canada's architects and then consciously created. We see a Canada which imagined itself into existence.[38]

The construction of the CPR is thus romanticized to communicate a technological nation that "equates the construction of the CPR with the constitution of Canada and praises each with reference to the other. Canada is valorized as a nation because it is the product of a technological achievement, and the railroad is the great product of heroic individuals who dreamt a nation."[39]

Literary approaches to nation-building reinforce this technologically oriented understanding, adding to the discussion an understanding of the fact that nation-building was environmentally destructive and destroyed Indigenous societies and cultures. Underscoring such critiques of nation-building are notions relating to air. Air is polluted with industrial emissions and through fossil fuels burning, and air is polluted with diseases such as tuberculosis, which is a primary factor in the immense reduction of Indigenous populations in western Canada after 1870.

Dirty Air, Nation-Building, and Literary Culture

Birk Sproxton's 1997 novel, *The Red-Headed Woman with the Black Black Heart*, is a good entry point to the *dirty air* thesis of this discussion. Sproxton was a Manitoba and Alberta poet and novelist who was born in 1943 in the northern Manitoba town of Flin Flon, where the novel is set, and who died in 2007. The novel's plot relates to the 1934 mining strike that shut down the mine and its smelter. The main character is a woman named Mickey Marlowe who is seen as either—depending on one's persuasion—a labor activist or communist criminal and foreign interlocuter who catalyzes civil disobedience and harms productive industrial activity. She is, in any case, an enemy of the crown and thus the primary forces of imperialistic and colonially derived activity in Canada. The extraction of raw materials and the industrial activities related to them are clearly seen as instruments of government, as Sproxton's narrator points out. The premier of Manitoba, for example, opened "up this town … ensuring a flow of $27 million into the plant construction—no small matter that, in 1928."[40] Catalyzing mining activity is the construction of a network of railways that stem from the trans-Canadian CPR line, thus connecting mining to the greater ambitions of nation creation. As much as industrial activities are themselves contributors to a national economy rooted in environmental degradation, exploitation of resources, control of territory, and displacement and control of Indigenous people, these activities are traced to the creation of the coast-to-coast railroad. As Sproxton's narrator asserts, "the geography" of Mickey's world is underscored by the workings of the railroad:

> Mickey lives among the web of the railroad. The main line stretches westward and from it once line branches off and runs north to Hudson Bay Junction. From the Junction one line runs north to Prince Albert and the other branches back into Manitoba where it makes another fork at The Pas. One great long line runs to Churchill and the waters of Hudson Bay. The other wends northward along the border with Saskatchewan into the Precambrian Shield. This place became crucial to Mickey's fate, the Manitoba Mining District.[41]

The story of Mickey Marlowe, Viennese-born disturber of the peace in northern Manitoba, points to the polluted air of Flin Flon, which is described as "the acrid odour of the smelting operations," a kind of "dragon smoke" that "bloomed and billowed,"[42] and is visible to all who arrive by train, as Mickey does.

Environmental degradation of the mining operations extends to the health of the town's inhabitants, as Mickey points out: "At home your kids are puking their guts out and running every fifteen minutes because of the stinking water they try to feed us. ... It's hard when your guts are heaving and you just want to get dry and keep your kids from getting sick."[43] The smokestack itself is described as "an obscene construction."[44]

Fellow prairie West author Robert Kroetsch's poems similarly intricately engage with themes of colonization, environmental degradation, and their relation to nation-building. Born in 1927 in Heisler, Alberta, in what is known as the "Parkland" in the center of the province, Kroetsch wrote over twenty novels, books of poetry, and non-fiction titles, and won the Governor General's Awards for Literature multiple times before his death in 2011. Kroetsch's interest in the relationship of colonial settlement to the land is evident in his 1974 poem "Stone Hammer Poem" and its exploration of a field stone that he keeps on his desk while he writes.[45] The stone serves multiple functions, especially signaling the connection of the speaker-poet figure to the landscape that he originates in and its intricate histories. As he writes, the stone

> smelling a little of cut
> grass or maybe even of
> ripening wheat or of
> buffalo blood hot
> in the dying sun[46]

Invoked in this passage is a span of time that encompasses settlement and farming and a time prior to the twentieth century when the plains of the Canadian prairies were abundant with buffalo (in fact, bison) who comprised an important part of the grasslands ecosystem. Eradicated through settlement and hunting, the bison of western Canada largely disappeared. Settlement and agricultural development caused their natural habitat to disappear. There are currently only about 30 percent of the original grasslands still in existence, and only small managed herds of bison exist (e.g., no wild herds). In Alberta, "more than 75 per cent of the province's original mixed grassland has been converted to tame forage (cultivated fields) and annual cropping. Only 17 per cent of its original fescue grassland remains."[47] With the disappearance of the bison, Indigenous peoples suffered owing to the loss of a primary food source.

For Kroetsch, the stone is literally and figuratively attached to the history of colonization, settlement, and agricultural development in Alberta, and the prairie West more generally. The stone is as

> old as the last
> Ice Age, the

retreating/the
recreating ice,
the retreating
buffalo, the
retreating Indians[48]

Inherently connected to the long span of history that culminates in the speaker's position as poet-author, he cannot escape its connection to the fraught forces of colonialism. The stone is found initially, as he writes, by his grandfather, in a landscape that has contested ownership:

This stone maul
was found.

In the field
my grandfather
thought was his

my father
thought was his[49]

Yet when the speaker inherited the land, the status of ownership is signaled as perilously attached to imperialistic forces of occupation. He sells the land to a man "who did not/notice the land / did not belong" to a host of previous occupants, including Indigenous peoples, "the Queen"—who represents the primary colonial force of Canada's West—and the CPR.[50] The CPR, as noted earlier, is a primary force in incorporating the West into the geopolitical space that would become Canada. Thus, the dirty air of the machinated transportation is a primary contributor to the processes of land occupation, agricultural development, and environmental and human degradation that Kroetsch signals with his poem about a simple field stone.

Similarly constructed around the concept of an object that catalyzes the poet's imaginative exploration of identity and place, "The Ledger," published in 1975, grows out of the discovery of a ledger that recounts the financial accounts of an ancestral water mill in Bruce County, Ontario. Serving as an attempt to account for the earliest settlement of his family members in Canada, "The Ledger" engages in a similar consideration of the environmental degradation that is connected to colonial and imperialistic enterprise. Quoting from the *Canada Gazette* of August 17, 1854, Kroetsch's speaker-poet reflects on the governmental discourse of the organization of land previously unclaimed by Europeans. The *Canada Gazette*, published since 1841, publishes "new statutes and regulations, proposed regulations, decisions of administrative boards, private sector notices and an assortment of government notices which are required by statute to be published so as to disseminate this information to the public":[51] "Notice is hereby given that the undermentioned lands ... in the County of Bruce, U.C., will be open for sale to actual settlers ... The price to be Ten shillings per acre ... Actual occupation to be Immediate and continuous."[52]

The purpose of such occupation is the development of land for the purposes of commerce that can only occur as a process of degradation. In the poem, Kroetsch points to a series of relationships that signal the degrading effects of European-colonial oriented occupation of the land. "To raise a barn," for example, results in cutting "down a forest," or "To raise oats and hay" means to "burn the soil." Similarly, "To raise cattle and hogs" is to kill a variety of wild animals, such as bear, mink, marten, lynx, fisher, beaver, and moose.[53] The scope of this work as colonial enterprise whose only purpose is to exploit land in aid of commercial development is accounted for in an unattributed passage that outlines a successful settler:

> a specimen of the self-made men who have made Canada what it is, and of which no section has brought forth more or better representatives than the County of Bruce. Mr. Miller was never an office-seeker, but devoted himself strictly and energetically to the pursuit of his private business, and on his death was the owner of a very large and valuable property.[54]

The phrase "self-made men" is repeated to emphasize the nature of nation-building enterprise and to point to the irony that the exploitation of the land, and its preexisting inhabitants, is less "self-made" than it relies on using preexisting features of geography and space. The water mill at the center of the poem, for example, serves symbolically to illustrate the exploitive relationship of nation-building to natural resources for the river is preexisting. The construction of the mill serves to harness the power of the river, both containing it and using it for industrially related purposes. Further, the mill's work links to the broader goals of shifting the purposes of the geography, allowing wood to be sawn and shaped for construction, as well as other mill-related activities.

Kroetsch and Sproxton, as well as Brandt and McKay, reflect primarily European-settler perspectives, even if Kroetsch can be seen especially to be attuned to Indigenous-related issues. I want to turn for the remainder of my textual discussion to two Indigenous people who write not directly in response to European positions but rather to their own perspectives as First Nations peoples who live alongside the encompassing enterprise of the Canadian nation and its development.

The Canadian Nation and Indigenous Experience

James Daschuk, in his influential study *Clearing the Plains: Disease, Politics of Starvation, and the Loss of Aboriginal Life*,[55] notes that the long history of diseases such as smallpox affecting health, population numbers, and social and communal organization within Indigenous people in Canada's West underscores the particularly deleterious effects of the creation of the Canadian nation on Indigenous people. He writes that "[t]he annexation of the north-west by the Dominion of Canada in 1870 changed the political, economic, and medical history of the region forever."[56] Tuberculosis became "the primary cause of sickness and death. ... and cut down the Indigenous population. An epidemic unlike anything the region had ever seen, it swept through the entire newly imposed reserve system."[57] Ironically, as mechanized transportation increased population movement across the West, and throughout the nation, the Canadian government invoked systems through which to limit the movements of Indigenous people, exercising control over them and also creating social conditions that promoted the spread of disease as well as restricting the options available to communities to evade the health disaster they were facing. For example, by 1883, as the CPR was nearing completion, almost all Indigenous people lived on reserves "under the control of government officials." These people were dependent upon "rations supplied by the government."[58] The railway is closely linked to the degradation of Indigenous populations, as Daschuk remarks: "The synergy between pre-existing sickness, hunger, and the spread of contagious diseases such as measles along the improved transportation system based on railway travel increased mortality on reserve populations that were bearing massive disease loads."[59]

The development of the railway is directly linked to the movement of Indigenous populations onto reserves, allowing not only for massive intensification of agriculture, but also for the control of Indigenous people. John A. MacDonald, prime minister of Canada in the early 1880s, employed a system of starvation to force people onto reserves with the promise of food that was as effective as it was cruel.[60] Such control facilitated the construction of the national railway. As Daschuk notes,

> [w]ith the exception of a few brief confrontations, construction of the Canadian Pacific Railway west of Swift Current continued almost unabated. By August 1883, the first train reached Calgary, linking the western plains to eastern Canada and the world. ... The railhead at Calgary marked completion of the infrastructure required for full-scale completion of the prairies.[61]

Two contemporary Indigenous authors, Joan Crate and Randy Lundy, through their varied and nuanced corpus of poetry and fiction, interlace the long-term effects of the development of the Canadian nation on Indigenous people. Writing out of a self-identified position as a Métis woman with Cree and other ancestry, Joan Crate addresses the challenges of Indigenous identity in an urban Indigenous world. Crate was born in Yellowknife in 1953 and identifies as a Métis woman of Cree and mixed European ancestry. Urban Indigenous identity is rooted in the genocidal tendencies of the colonial and Canadian forces that marginalized and silenced Indigenous people. In her poem "Unmarked Graves" from the "Loose Feathers on Stone: For Shawnandithit" sequence in her 2001 collection *Foreign Homes*,[62] Crate articulates a shared history with Shawnandithit and the Beothuk. The Beothuk were a First Nations tribe of Newfoundland who were eradicated through genocide and disease brought by Europeans, and Shawnandithit was the last survivor before succumbing to disease. In the opening lines to the poem, Crate's speaker registers her affinity with the Beothuk and the weight of colonial history, noting that "There is no stone, no work or prayer to mark / Our fleet lives, our staggering deaths."[63] History is burdened by the racist actions of white settlers and colonial administrators: "Everything / We were is buried in silence under dark/ White plots,"[64] which accompany "Our bullet and virus bones."[65] The past and future are enmeshed in the notion of progress that is measured as "contempt and new subdivisions,"[66] signaling the relation of colonial land acquisition through force and the development of contemporary housing in the cities and towns on occupied space.

This consideration of the contemporary and the historical in the lineage of current Indigenous existence in urban settings is placed alongside the poet's lyric considerations of the lost people in her life, in the poem "You have disappeared." These people are lost to drug addiction and alcoholism, and "pock my dreams with disease,"[67] as she states. The word "pock" reinforces the connection of contemporary addiction to the historical role of tuberculosis to eradicate Indigenous people. The figure of the lost and broken Indigenous person and effects of colonially rooted dysfunction is reinforced in the portrait of the speaker's former husband in "Dirty Dream." The broken man is contrasted with his younger self where he was "young and sharp / … at your best," a figure marked by his "long black hair, perfect teeth."[68] Encountering him years after their breakup, the speaker finds a man diminished: "your son and I found you on the street, / your mouth grimacing recognition."[69] The former husband's existence is marked by the metaphor of disease, and the wound she recognizes him as figures the greater wound of colonialism and its ongoing effects:

> I wrapped my arms around you and for an instant
> loved you
> not for what you were or could have been,
> did or might have done,
> but for the open wound you are.[70]

This sense of woundedness runs through Randy Lundy's 2018 collection *Blackbird Song*.[71] Lundy is a member of the Barren Lands Cree Nation of northern Manitoba, where he was born. He has lived in Manitoba and Saskatchewan, and has lived in Toronto since 2020 where he is a professor of creative writing at the University of Toronto. His poems range through considerations of ancestry and place, and acts of memory, as they attempt to grapple with the weight of the forces that shape his existence. He situates his speaker often at margins as he considers memory and self. These margins are marked by the signs of nation-building as his speaker is often positioned in a space that is edged by the CPR mainline on one side and fields containing howling coyotes on another. Occasionally, the space is identified as the rural hamlet of Pense, about 20 km to the west of the city of Regina. In a poem titled "For Kohkum, Reta," Lundy opens with a conventional lyric moment. *Khokum* is the Cree word for Grandmother. He writes:

> Might as well settle
> in a hard-heavy chair,
> with a mug of steaming
> green tea and some Powwow or
> plainchant turned up loud.
> Stare out the window
> into the glare of sun glazing
> the snowfield south of the house.[72]

Lundy frames the moment through the hybrid cultural influences of Indigenous powwow, a form of music that is informed by a drumbeat and is communal, involving multiple participants, and plainchant, a form of liturgical chant in a Christian tradition. Similarly, the space he occupies is marked by cultural duality as "a mile north of here,/ a train engine will groan and then roar,/ a biblical beast, a wihtikow come for the feast."[73] In Indigenous culture, the wihtikow (also known as a wendigo) is "a malevolent, malicious entity, a cannibal and also as a supernatural (manitou) possessing a great spiritual force that lives preferably in the forests."[74] The combination of wihtikow and train engine reinforces my thesis of the relationship of transportation technology to the subjugation of Indigenous people, for the wihtikow is a consumer of humans, leading only to death and destruction.

In another poem, "Midnight, Early Spring," similarly situated in the speaker-poet's home, memory of ancestry is summoned to consider the fraught nature of the poet's childhood, where a positive memory of being with his grandmother is juxtaposed with the absence of his mother in his youth:

> What was it you were thinking? You were thinking about remembering.
> Remembering a small brown-skinned boy in the summer sun in a large
> opening among the pine trees, his grandmother's eyes glowing with
> pride for no other reason other than that he was alive, holding in his
>   small
> hands a jute back of seed for her garden. The order there. The
> straight lines. The love.[75]

But as he states, "By this time / your mother was lost to you for the next twenty years."[76] And what is this loss marked by? Similarly to the wounded ex-husband in Crate's poem, women such as the poet-speaker's mother were victimized by the men who "came from across the continent/ to the town you grew up in":[77]

> Women in this town
> bought their groceries,
> sent their children to school,
> washed their floors
> always with their eyes lying
> in dirty pools at their feet.
> They knew where the predators were.
> They went to bed with them
> every night.[78]

The image in Lundy's poem of the invading man, predatorial and destructive, serves to underscore the fundamentals of transportation technology and nation-building, which are bound up in the forces of colonialism. Indigenous people are seen as resources that can be drawn on, despite the negative effects of their interactions, in a similar way to how natural resources are extracted from the landscape. The process of nation-building is made possible by technological advancements in transportation, which inherently rely on the burning of fossil fuels, thus producing dirty air, and likewise dirtying the "air" metaphorically. The dirty coal-burning engines of the nineteenth century are supplanted by the diesel fumes of more recent trains, and the smelter smoke is replaced by the environmental destruction of Canada's oil sands. Oil sands production produces numerous environmental effects, including contamination of food sources that Indigenous people use,[79] and contamination of major waterways such as the Athabasca and Mackenzie rivers in Alberta and the Northwest Territories.[80]

Conclusion

Transportation technology, then, is employed in the service of a commodified economy that relies upon the repurposing of land from wilderness to European-based agricultural and mineral use. Long-standing inhabitants of this landscape are displaced and contained, subjugated to the forces of nation-building, their air polluted by fossil fuels and by transmittable diseases. This process in turn led to political action that disempowered and controlled Indigenous people. Challenges to these processes are ongoing, including contested application of the term "genocide" to denote the mass destruction of Indigenous populations in the Americas; nevertheless, scholarly estimates put the number killed or prematurely dead in the western hemisphere since European contact at 100 million.[81] In Canada, the horrors of the residential school program, which forcibly removed children from their parents to reeducate and assimilate them, are becoming increasingly clear. The program started in 1894 and continued until the last school was closed in 1997. In an ongoing process, unmarked graves of children are being uncovered—at current count upwards of 1,000 graves have been identified, but many more are expected.[82]

Dirty air, as we have seen, reflects metaphorically in a corpus of Canadian literary expression the larger processes of nation-building that has led to a conflicted nation. The strength of the metaphor of dirty air to articulate the destructive nature of nation-building is well captured by poet and novelist Daphne Marlatt in her long prose poem *Steveston*. She addresses the myriad industrial forces that have reshaped the landscape and people who inhabit the Fraser River delta town of Steveston, British Columbia. The development of the town is marked by the forces that shape a nation: "This corporate growth that monopolizes the sun, moon & tide,"[83] she writes, offering a critique of the industrial nature of the building of a nation. Such a nation is imbued with "the trucks of production."[84] The boats that frequent the Fraser River burn fuel, literally marking their presence by the smoke they give off, such as "The chugging of an Easthope moving east to west" (an Easthope is a type of engine used in boats) that penetrates her awareness as she considers the river in front of her.[85] Every sign of nation that Marlatt encounters in Steveston signals the negativity of nation-building. Canada is "a vision of telephone poles, wires, cement." In Steveston, this is rooted in the exploitation of salmon expressed as the "Dream of seizing silver wealth that swims, & fixing it in solid ground, land … A mis-reading of the river's push."[86] The ultimate conclusion to this misreading is repurposing of land to contain humans, always a negative force, as Marlatt notes: "waiting the bulldozer tracts, demolition, scaffolding & sewer drains, the ript denuded soil patches of artificial grass will cover, like burial plots."[87]

For Marlatt, as for fellow Canadian authors McKay, Brandt, Sproxton, Kroetsch, Lundy, and Crate the story of the destructive building of a nation is made possible by the pollution created by the burning of fossil fuels—by the creation of dirty air. Inhabitants of the Canadian nation become "citizens of an exploited earth."[88] This is in contrast to Marlatt's vision for the town of Steveston that is in concert with natural cycles: "Steveston: delta, mouth of the Fraser where the river empties, sandbank after sandbank, into a muddy gulf."[89] The nation, however, destroys such an organic vision for space—"We obscure it with what we pour on these waters, fuel, paint."[90] The nation is thus constructed out of "dirty air"; these writers attest that the metaphor of dirty air has the power to figure a nation conflicted in its relation to its landscape and people, leaving both in a fraught state.

Notes

1   Kenneth Pletcher, "Trans-Canada Highway," *Britannica*, https://www.britannica.com/topic/Trans-Canada-Highway, accessed May 4, 2023.

2   https://www.thecanadianencyclopedia.ca/en/article/railway-history, accessed May 31, 2023.

3   https://www.britannica.com/topic/Trans-Canada-Highway, accessed May 31, 2023

4   https://www.rcaanc-cirnac.gc.ca/eng/1100100013785/1529102490303, accessed
     May 23, 2023.

5   https://www.rcaanc-cirnac.gc.ca/eng/1450124405592/1529106060525, accessed May
     31, 2023.

6   Birk Sproxton, *The Red-Headed Woman with the Black Black Heart* (Winnipeg:
     Turnstone Press, 1997).

7   Jonathan Naylor, "When the Smoke Stopped: The Shutdown of the Flin Flon
     Smelter," *The Reminder*, February 20, 2017, https://www.thereminder.ca/local-news/
     when-the-smoke-stopped-the-shutdown-of-the-flin-flon-smelter-4103552, accessed
     May 31, 2023.

8   Naylor, "When the Smoke Stopped."

9   https://flinflonheritageproject.com/mining/wppaspec/oc1/lnen/cv0/ab204, accessed
     May 31, 2023.

10   Max Liboiron, *Pollution is Colonialism* (Durham, NC: Duke University Press, 2021),
     5–6. https://doi.org/10.1515/9781478021445

11   Glenn Sean Coulthard, *Red Skin, White Masks: Rejecting the Colonial Politics of
     Recognition* (Minneapolis: University of Minnesota Press, 2014), 7. https://doi.
     org/10.5749/minnesota/9780816679645.001.0001

12   Jen Bagelman and Sarah Marie Wiebe, "Intimacies of Global Toxins: Exposure &
     Resistance in 'Chemical Valley,'" *Political Geography* 60 (2017), 76. https://doi.
     org/10.1016/j.polgeo.2017.04.007

13   Bagelman and Wiebe, "Intimacies of Global Toxins," 77.

14   Bagleman and Wiebe, "Intimacies of Global Toxins," 82.

15   C.N. Westman and T.L. Joly, "Oil Sands Extraction in Alberta, Canada: A Review
     of Impacts and Processes concerning Indigenous Peoples," *Human Ecology* 47 (2019),
     233–43. https://doi.org/10.1007/s10745-019-0059-6

16   Lena Gross, "Fuelling Toxic Relations: Oil Sands and Settler Colonialism in Canada,"
     Anthropology Today 37, No. 4 (2021), 19–22. https://doi.org/10.1111/1467-8322.12666

17   Gross, "Fuelling Toxic Relations," 22.

18   Di Brandt, *Now You Care* (Toronto: Coach House Books, 2003).

19   Brandt, *Now You Care*, 31.

20   Brandt, *Now You Care*, 34.

21   Don McKay, *Apparatus* (Toronto: McClelland & Stewart, 1997).

22   McKay, *Apparatus*, 53.

23   McKay, *Apparatus*, 53.

24   McKay, *Apparatus*, 53.

25   McKay, *Apparatus*, 53.

26   McKay, *Apparatus*, 53.

27   McKay, *Apparatus*, 57.

28   McKay, *Apparatus*, 56.

29   McKay, *Apparatus*, 58.

30   Ben Bradley, Jay Young, and Colin M. Coates, "Moving Natures in Canadian History:
     An Introduction," in *Moving Natures: Mobility and Environment in Canadian History*,
     ed. Ben Bradley, Jay Young, and Colin M. Coates (Calgary: University of Calgary
     Press, 2016), 2. https://doi.org/10.2307/j.ctv6cfr6m

31 Bradley, Young, and Coates, "Moving Natures in Canadian History," 2.
32 Bradley, Young, and Coates, "Moving Natures in Canadian History," 3.
33 Bradley, Young, and Coates, "Moving Natures in Canadian History," 3.
34 Bradley, Young, and Coates, "Moving Natures in Canadian History," 4.
35 Bradley, Young, and Coates, "Moving Natures in Canadian History," 5.
36 Bradley, Young, and Coates, "Moving Natures in Canadian History," 5.
37 Bradley, Young, and Coates, "Moving Natures in Canadian History," 6.
38 Maurice Charland, "Technological Nationalism," *Canadian Journal of Political and Social Theory* X, no. 1–2 (1986): 196.
39 Charland, "Technological Nationalism," 196.
40 Sproxton, *The Red-Headed Woman with the Black Black Heart*, 3.
41 Sproxton, *The Red-Headed Woman*, 3.
42 Sproxton, *The Red-Headed Woman*, 23.
43 Sproxton, *The Red-Headed Woman*, 89.
44 Sproxton, *The Red-Headed Woman*, 105.
45 Robert Kroetsch, *Completed Field Notes: The Long Poems of Robert Kroetsch* (Toronto: University of Alberta Press, 2000).
46 Kroetsch, *Completed Field Notes*.
47 https://Www.Natureconservancy.ca/En/What-We-Do/Resource-Centre/Conservation-101/Grasslands.Html#:~:Text=In%20Alberta%2C%, accessed May 31, 2023.
48 Kroetsch, *Completed Field Notes*, 4.
49 Kroetsch, *Completed Field Notes*, 4.
50 Kroetsch, *Completed Field Notes*, 6.
51 https://publications.gc.ca/collections/collection_2014/gazette/SP7-1-2001.pdf, 18, accessed May 31, 2023.
52 Kroetsch, *Completed Field Notes*, 13.
53 Kroetsch, *Completed Field Notes*, 13.
54 Kroetsch, *Completed Field Notes*, 15.
55 James W. Daschuk, *Clearing the Plains: Disease, Politics of Starvation, and the Loss of Aboriginal Life*, Canadian Plains Studies 65 (Regina: University of Regina Press, 2013).
56 Daschuk, *Clearing the Plains*, 25.
57 Daschuk, *Clearing the Plains*, 25.
58 Daschuk, *Clearing the Plains*, 28.
59 Daschuk, *Clearing the Plains*, 30.
60 Daschuk, *Clearing the Plains*, 207.
61 Daschuk, *Clearing the Plains*, 207.
62 Joan Crate, *Foreign Homes* (London, ON: Brick Books, 2001).
63 Crate, *Foreign Homes*, 45.
64 Crate, *Foreign Homes*, 45.
65 Crate, *Foreign Homes*, 45.
66 Crate, *Foreign Homes*, 45.
67 Crate, *Foreign Homes*, 29.
68 Crate, *Foreign Homes*, 29.
69 Crate, *Foreign Homes*, 29.
70 Crate, *Foreign Homes*, 29.

71 Randy Lundy, *Blackbird Song*, Oskana Poetry & Poetics (Regina: University of Regina Press, 2018).

72 Lundy, *Blackbird Song*, 6.

73 Lundy, *Blackbird Song*, 6.

74 https://Virily.Com/Culture/Legendary-Creatures-of-Canadas-First-Nations-the-Wendigo/, accessed May 31, 2023.

75 Lundy, *Blackbird Song*, 37.

76 Lundy, *Blackbird Song*, 37.

77 Lundy, *Blackbird Song*, 72.

78 Lundy, *Blackbird Song*, 72.

79 Abha Parajulee and Frank Wania, "Evaluating Officially Reported Polycyclic Aromatic Hydrocarbon Emissions in the Athabasca Oil Sands Region with a Multimedia Fate Model," *Proceedings of the National Academy of Sciences* 111, no. 9 (March 4, 2014): 3344–49. https://doi.org/10.1073/pnas.1319780111

80 David W. Schindler, "Unravelling the Complexity of Pollution by the Oil Sands Industry," *Proceedings of the National Academy of Sciences* 111, no. 9 (March 4, 2014): 3209–10. https://doi.org/10.1073/pnas.1400511111

81 "David E Stannard, *American Holocaust: The Conquest of the New World* (New York: Oxford University Press, 1992).

82 https://www.culturalsurvival.org/news/new-revelations-child-graves-residential-schools-lays-bare-history-genocide, accessed May 31, 2023.

83 Daphne Marlatt and Robert Minden, *Steveston* (Vancouver: Ronsdale Press, 2001), 18.

84 Marlatt and Minden, *Steveston*, 11.

85 Marlatt and Minden, *Steveston*, 29.

86 Marlatt and Minden, *Steveston*, 30.

87 Marlatt and Minden, *Steveston*, 38.

88 Marlatt and Minden, *Steveston*, 52.

89 Marlatt and Minden, *Steveston*, 56.

90 Marlatt and Minden, *Steveston*, 57.

3

Witnessing *Challenger*: Viewing Aerial
Space through the Reverberations
of Disaster

Chantelle Mitchell

In 1986, a catastrophe occurred in the skies over Cape Canaveral, Florida, simultaneously broadcast across the United States, witnessed by countless many. This chapter turns upon the image of this specific disaster, the explosion of the space shuttle *Challenger*, in order to read the ripples of the event across and into the contemporary. In taking this space shuttle explosion as the object of focus, detouring through contemporary philosophy, understandings of aerial connectivity, and transformations of aerial space, the event is seen anew in the present. This viewing opens up possibilities for aerial revelation, moving from a discrete, localized aerial disaster toward a continually unfolding aerial disaster in the present, that of increasing carbon dioxide (CO_2) levels and the ongoing climate crisis. Beginning with considerations of air and atmosphere, this text locates and historicizes the *Challenger* disaster, bringing together multiple understandings across literature, in particular the reflections and suppositions of Paul Virilio and Michel Serres, before encountering the *Challenger* explosion as an aerial disaster. In understandings of the power and influence of views from and of the air, revelations for relationships to the air itself can emerge. In attending to the ripples of this aerial disaster, the complexity of the airspace as lived and perceived space emerges, opening the framework of the disaster in recognition of shifting relations to aerial space. Following this path through the cloud of the explosion does not present strategies for the mitigation of negative aerial conditions, but moves toward understandings of the power of events and their representation in the recognition of air as a complex, inhabited site of multiplicity and connectivity surrounding the Earth.

Chantelle Mitchell, "Witnessing *Challenger*: Viewing Aerial Space through the Reverberations of Disaster" in: *Imagining Air: Cultural Axiology and the Politics of Invisibility.* University of Exeter Press (2023). © Chantelle Mitchell. DOI: 10.47788/OTME3403

This chapter presents a speculative reading of the *Challenger* space shuttle explosion as a significant event that illuminates (literally and figuratively) aerial space as a material, social, and political site. Bruno Latour, in the context of a broader political theorization for an exhibition in 2005, presented the entangled concept of *Dingpolitik* within a catalogue essay. Nestled within historical detours and future speculations was a reference to the 2001 *Columbia* space shuttle disaster alongside the corresponding image of a NASA hangar filled with collected debris.[1] This debris was laid out to mimic its placement within the shuttle before the catastrophe struck. Referencing this image, Latour terms this assembly and its representation an "exploded view" of the shuttle, in which the complicated entanglements of material and other factors was accumulated.[2] In a similar manner, this text attempts an exploded view of the 1986 space shuttle *Challenger* disaster, with reference to the iconic image of the shuttle's explosion against the backdrop of a clear blue sky (see Fig. 3.1), and the aerial significance of the accident. Understanding how an event such as *Challenger*, suspended as it was between the Earth and the cosmos, reads as significant nearly forty years following its explosion, and in the context of air, necessitates an exploded view. In doing so, reverberations of the disaster are traced across disaster studies, contemporary philosophy, and new materialist and political entanglements, presenting an exploded view of the *Challenger* disaster in the context of relationships to air.

Aerial Catastrophe Par Excellence

On January 28, 1986, at NASA's Kennedy Space Center Launch Complex in Cape Canaveral, Florida, the *Challenger* Space Shuttle was launched. This launch was long awaited: Originally scheduled for July 1985, it was pushed back to November, before being rescheduled for January 1986. The original January date, the 23rd, quickly slipped to the 24th, 25th, and so on, owing to a myriad of factors including launch pads and inclement weather. On January 28, however, onboard were seven crew members, intending to launch satellites, study comets, and, as part of the Teacher in Space Project, allow non-astronaut civilian Christa McAuliffe to teach lessons from space. This launch attracted significant attention because of McAuliffe's civilian status and future possibilities of civilian space presence. The night before the shuttle's launch, a cold front swept across the southernmost parts of Florida, freezing parts of *Challenger*'s external structure. Despite combined factors, including the drastic overnight drop in temperature and days of flight cancellations, the shores of Cape Canaveral filled with spectators. These included the families of the crew members, some of McAuliffe's students, and media outlets, all to witness the launch against the backdrop of a cloudless blue sky. The events that followed have been inscribed in cultural memory, continuing to reverberate into the present.[3]

Seventy-three seconds after *Challenger*'s launch, a critical failure saw the space shuttle dramatically explode, sending smoke and hydrogen into the skies. This catastrophic event tragically killed all onboard and spread debris into the waters off Cape Canaveral, some of which continued to rain down into the ocean for forty-five minutes following the initial explosion.[4] This shocking incident was not only witnessed by audiences present at the launch site, but was also experienced across the United States as television networks aired taped replays of the catastrophe.[5] Schools across the country broadcast live feeds of the event in classrooms, in celebration of McAuliffe's expected journey, amplifying the emotional disturbance of the tragedy. The explosion of the space shuttle was an event that played out across the media, garnering world-wide attention, sparking inquiry, inquests, and drawing due attention to cause. The mediatization of the event aligns with the notion of the "event instant," as presented by Virilio, a unique dissemination of the event through the medium of television. *Challenger*'s explosion was the first "in-air" loss of life for the United States' space program, an event that reverberated as collective disbelief and national grief.[6]

This chapter does not seek to retread the causal grounds of this disaster, but rather looks toward an interrogation of the ripples of this explosion culturally and philosophically, thinking through air as connective medium. On the evening of January 28, 1986, President Ronald Reagan was due to give the State of the Union address, mere hours after the planned ascent of the shuttle. Some believed that NASA's refusal to delay the launch for warmer weather was tied to this impending event, but regardless, instead of presenting his prepared speech, for the first time in the nation's history the State of the Union was delayed, with Reagan addressing the nation in memoriam, as opposed to the planned celebration. Reading from the World War II pilot John Gillespie Magee's poem "High Flight," Reagan mourned the loss of the seven aboard the space shuttle, positioning them as slipping from the "surly bonds of Earth."[7] However, thinking materially and aerially, the *Challenger* disaster was an earthly catastrophe—with the shuttle never breaching the outer atmosphere and never crossing into the frontier of space. In taking place as it did, in the clear blue skies of Florida, and in the intense mediatized landscape of its witnessing, this disaster can be positioned as worldly. The *Challenger* space shuttle was to be the twenty-fifth space shuttle launch into space, an event quickly turning into routine, but became cemented in historical and cultural memory through a coalescence of factors, notwithstanding the vision of disaster.

The *Challenger* disaster became a shared event through broadcast, which would quickly filter across the globe. On the eve of the catastrophe, Reagan acknowledged: "We will never forget them, nor the last time we saw them," and it is the seeing that is significant.[8] Radio broadcaster Rick Van Cise, in reflecting upon the disaster, acknowledged the catastrophe as "a visual story," his memories of the day landing upon the color of the smoke from lift-off to explosion, changing from sooted black to pure white.[9] The imagery of the catastrophe appears as visually striking: White clouds billow in multiple directions, markers of velocity and explosion, set against the backdrop of a

Fig. 3.1. Booster plume and expanding ball of gas: the space shuttle *Challenger* explosion, January 28, 1986. Image credit: NASA.

clear, vibrant blue sky. A perfect backdrop, the empty clear skies of Cape Canaveral, became vessel and medium through which the catastrophe of the *Challenger* disaster can be apprehended. There is an eerie beauty to this image—decontextualized, it may appear akin to one of the more tempestuous cloud studies of painter John Constable—billowing plumes in motion, against the backdrop of a blue sky. His *Study of Cirrus Clouds* from 1882 has no tree line or human presence—just the capture of clouds moved by the wind across the sky, organic forms that mirror the catastrophic plumes emanating from the shuttle.[10]

Material and Conceptual Matters of Air

The explosion of the *Challenger* space shuttle in 1986 was a catastrophe that filled television screens and courted significant political attention in the weeks and months following the incident. Positioned as the "worst space disaster" in history and the "first fatal in-air accident" in US history, the event was undoubtedly significant; but it is the "in-air" qualifier and the aerial qualities of this disaster that are particularly crucial for this investigation.[11] In fore-grounding considerations of the nature of the *Challenger* explosion in this way, the air is identified as the site of the disaster. The aerial setting of the explosion shapes perceptions of the incident, while the incident illuminates certain material, social, and political qualities of the air—qualities that are often nonvisible or obscured. Understanding the *Challenger* disaster as occurring within the air, and documented visually as an aerial disaster, necessitates understanding air's complex materiality, which illuminates the significance of the disaster more broadly in material and socio-political frames.

Architect and researcher Nerea Calvillo recognizes air's multiplicity, defining air as "a space, an object, a threat, a myth, a weapon, a common."[12] Amid this multiplicity, there is also the important condition of air's perceived absence. In meditating on air's material complexity, literary scholar Steven Connor argues that air "encompasses its own negation, indeed perhaps negativity itself."[13] He demonstrates this by stating that even if air itself is taken away, "the empty space you have left still seems to retain most of the qualities of air."[14] This perceived absence and negativity is illuminated in fundamental relationships to the air developed in childhood. Behavioral studies of drawings of children demonstrate this, with the absence of a true sky almost universal in illustrations of the landscape, revealing a common perception at the center of early understandings of air and the sky more broadly.[15] In many of the drawings analyzed, a gap between the horizon line and the sky itself exists—a shared experience that is explained by the development of three-dimensional awareness over time.[16] The absence of the sky is an absence of air, or the atmosphere, in representation—an empty space. Lacking the fixity of geology, the weightedness of machines, or the presence of plants and other sentient and nonsentient beings, air and the atmosphere are able to slip from the mind's eye. In studies of children's drawings of landscapes, the sky creeps closer to the ground as children age, the absence of the atmosphere gradually filling in with time. However, it can be argued that this perceived absence is a lasting apprehension: The empty space between the blue sky at the top of the page and the ground colored in at the bottom is representative of a pervasive and lasting misapprehension of the material nature of the world.

Air's ontological absence, even once acknowledged, is hard to shake. This is a condition that environmental humanities scholar Steve Mentz understands as contributing to the voidal, vacuum-like affective resonance of air.[17] Speaking against philosophical tradition, Luce Irigaray criticizes the historical "forgetting of air" in the Western philosophical tradition, which has transformed air into a conceptual void "by using up the air for telling without ever telling of air itself."[18] Furthering this, Irigaray states that "air does not show itself. As such, it escapes appearing as (a) being. It allows itself to be forgotten."[19] Tracing air's apprehension across time and referencing significant scientific developments in the study of air, Mentz notes that air is visually imperceptible: "The thin-ness of air, its resonance as a space into which things vanish, emerges from a cultural equivalent of the horror vacui ... Vacuums must be filled, emptiness cannot last: that's the lesson of the air pump and the barometer. But our eyes say otherwise."[20] The material condition of air is understood to be a condition complicated by ocularcentric preoccupations, with air commonly perceived as nothingness.

In addressing the history of the landscape as collective political and social construct, Kenneth R. Olwig identifies the significance of linear perspective and landscape painting in fixing perceptions of space.[21] Following Olwig, it becomes possible to rethink understandings of air as nothingness in the development of frameworks from which air's absent presence might be perceived. In seeking landscape's other half (and perhaps this is the empty space of children's landscape drawings), Olwig turns to the imagined discipline of *aerography*, which displaces the geologic of *geography* with air. Drawing from the historical concept of aether, Olwig quotes from Arthur Stanley Eddington, in an acknowledgement of aether and air that mirrors contemporary reflections.[22] Eddington writes:

> If you can make this reversal of the picture, turning space from a negative into a positive, so that it is no longer a mere back-ground against which the extension and the motion of matter is perceived but is as much a performer in the world drama as the matter is, then you have the gist of the aether theory whether you use the word "aether" or not.[23]

The negative condition of air gives rise to its social and cultural apprehensions, but it is when air becomes apparent that it can be perceived as a shared condition, whether that is through conditions that make air unbreathable or airborne vectors that lend it visibility. This visibility, the apparentness, can be linked to catastrophic climate crisis. As dust from the Saharan desert fills the skies of Europe or the smoke from the catastrophic 2019/2020 fire season in so-called Australia circumnavigates the globe, connectivity to and through air across borders becomes apparent.

In this chapter, the diffuse nature of air entangles with atmosphere. Both become difficult at times to distinguish, and common patterns of speech see slippage between the two. As Eva Horn reveals, the atmosphere is implicit within air, a composite part of air's structure. She notes that "Air is generally defined as the atmosphere of the earth, the layer of different types of gas surrounding the planet," within the air, the atmosphere.[24] The word "atmosphere" has a material, scientific definition—but alongside this, it has an affective component. A location, a work of literature, or of music might invoke an atmosphere, some *feeling* tied to the experience itself. Drawing from Derek P. McCormack's understanding of atmospheres, which he reads through the speculative framework of the balloon, atmospheres can be positioned as "elemental spacetimes that are simultaneously affective and meteorological, whose force and variation can be felt, sometimes only barely, in bodies of different kinds."[25] For philosopher Gernot Böhme, a key quality of atmospheres is that they are "suspended in the air."[26] Kathleen Stewart, with debt to noted affect theorist Lauren Berlant, attends to atmospheres through the notion of attunement, a means of perceiving the world within which "things hanging in the air are worth describing."[27] These affective atmospheres, which hang in the air and are encountered through complex relationships of human and more-than-human beings and elements, necessitate attention. In considering the criticality of air and atmosphere here, it is important to note the earliest definition of atmosphere is tethered not to the airs of the Earth, but rather an interstellar air. As historian Siobhan Carroll details, "Atmo-sphæra" was coined in 1638 by English writer John Wilkins, to describe what he imagined to be an "orbe of grosse vaporous aire" surrounding the moon.[28] Here, read against air as a crucial site for investigation, a similar speculative task is approached: Instead of a balloon, a monumental historical disaster is utilized as a carrier for inquiry into air's diffuse but crucial significance.

Unfolding and Unfurling "Disaster"

In the same way that air and atmosphere entangle, disaster, catastrophe, and accident find common ground. When a catastrophe happens, Michel Serres writes, "we say, 'it wasn't on purpose, it wasn't a sacrifice, but an accident,' inevitable, even calculable, through probabilities."[29] In June 2001, *The BMJ*, a journal published by the British Medical Association, banned the word "accident" from use in future articles, citing the imprecision of the term.[30] It was determined that the word "accident" obscures the nature of an event, or incident, with a motor-vehicle accident better described as a motor-vehicle collision, positing that much of what is commonly designated as accident has identifiable cause. In the same way that uses of air and atmosphere entangle and encounter equivocation, it can be seen that Serres' catastrophe, read as an accident, encounters further categorical complication in the context of contemporary disaster studies.

    The field of disaster studies is a complex landscape entangling risk management, policy, preparedness, and recovery.[31] Of interest in the context of this chapter are conceptual frameworks of the categorization of a disaster and how this might illuminate the *Challenger* disaster in a meaningful way. Critiquing the traditional field of disaster studies, theorists Jacob Remes and Andy Horowitz note that "Existing research [in the field of disaster studies] often assumes the category of disaster as an objective given and aspires to a technical analysis of achievements and failures."[32] This has been undertaken in the context of the *Challenger* disaster, most notably by Diane Vaughn.[33] However, Remes and Horowitz seek an alternative to the traditional field, outlining what they term to be critical disaster studies. This presents alternative modes of investigation against a more than 100-year lineage of traditional disaster studies. Significantly, it locates at its core the recognition that "disasters are interpretive fictions."[34] By interpretive fictions, it is meant that disasters are socially constructed, but also that they are ideas and events. The explosion of the *Challenger* space shuttle will be interrogated within this frame, drawing upon interdisciplinary sources that are identified as key to challenging traditional histories of the disaster. Another significant component of this approach is the recognition that "disasters are political," their experience, mediation, and response enfolded into structures of govern-mentality, power, and control.[35] This presents as crucial in understanding the *Challenger* disaster as a specifically aerial disaster amid contemporary contexts of airspace. However, before this is interrogated, the interpretive fiction of the *Challenger* disaster necessitates attention.

Just as critical disaster studies does not aspire to the "bullet-pointed knowledge of best practices" in traditional studies, neither does this chapter.[36] Rather, the complexity of the Challenger explosion as event entangles with broader complexities of the field. Within these, Franz Mauelshagen, a histo-rian engaged in climate research, highlights a lack of consensus across disciplines between the terms "catastrophe" and "disaster," exploring the semantic development of these terms over time.[37] This slippage reveals the reverberations of such events, but at times obscures their nature. In searching for the origin of the accident through industrial frames, Virilio presents, in arguably his most quoted statement:

> The accident is an inverted miracle, a secular miracle, a revelation. When you invent the ship, you also invent the shipwreck; when you invent the plane you also invent the plane crash; and when you invent electricity, you invent electrocution. ... Every technology carries its own negativity, which is invented at the same time as technical progress.[38]

With the "inverted miracle," the accident becomes a consequence of mechan-ical, industrial progress, tied as it is to catastrophe and disaster. In the collision of the negative space of the inverted miracle and that of the reversal of the image that becomes background, there is an explosion of presencing that reveals air and atmosphere as shared condition. Air's aforementioned negative condition aligns with what noted ecocritical and postcolonial studies scholar Elizabeth DeLoughrey understands about light—an absent presence with revelatory potential in uncovering relationships between people, ecolog-ical, and materiosocial, political entanglements within which such elements interrelate.[39] In considering aerial catastrophe through the lens of the nega-tive condition and the inverted miracle, it becomes possible to perceive collective responses as spreading through air understood as a connective medium. Indeed, disaster researchers Scott Gabriel Knowles and Zachary Loeb position the disaster as an "ordering mechanism," a specific event that is revelatory in the context of relationships to the world.[40]

In the current moment, lived experience of the COVID-19 pandemic has illuminated connectivity amid a specific biological disaster that itself is inex-tricably tied to air. COVID-19 presents the possibility of a spread traversing the globe either through droplets that spray like aerosols, linger in the air, and enter the body through respiration, or as a result of transit, which sees COVID-19 proliferate with the assistance of a global aerial transit network.[41] The rise in "contact tracing" through the spread of disease reveals air to be inextricably tied to people. Many public health responses to COVID-19 recommended gathering outdoors, lest the contained air of indoor space become filled with virus-causing droplets. While air might not immediately give rise to metaphors of conduction in the same way as water, air is a space within which bodies, ideas, and affects are transported, transmitted, and transformed. Returning again to DeLoughrey, she notes that light is an "invisible companion who accompanies us inwardly as much as it does outwardly," and through events such as the COVID-19 pandemic, it becomes apparent that air shares that same quality. [42]

In grappling with the event through the frames of cultural history, Gregory Whitehead reimagines the families of the *Challenger* astronauts on the morning of January 28, standing together upon a special viewing platform and looking toward the theater of the sky. Despite the daylight, Whitehead places these witnesses as tragic stargazers, perfectly positioned to take in the "necrodrama that followed the fire + fire + light."[43] Here, the catastrophic failure of the shuttle gives way to disaster—the etymologically "bad or evil star," a tragic portent in the clear blue skies above Cape Canaveral.[44] The field of disaster studies has been greatly shaped by the early and foundational contributions of scientist Charles Fritz, who proposed the definition of a disaster as "concentrated in time and space."[45] This temporal and spatial limitation positions a disaster or catastrophic event as necessarily sudden, or quick, which aligns with Virilio's conceptual framings. In an interview from 1998, he remarks: "The original industrial accidents as, for instance, the derailment of a train or the crash of an airplane, were all specific, localized, and particular accidents. They were taking place at a certain place and at a certain moment in time."[46] At first glance, the catastrophe that befell the space shuttle *Challenger* not only fits traditional categories of the disaster—temporally immediate, located in space and with great reverberations across society—but is also identified by Virilio as foundational. He writes: "The explosion of the *Challenger* space shuttle is a considerable event that reveals the original accident of the engine in the same way as the shipwreck of the first ocean liner."[47] As "original accident of the engine," the space shuttle is a revelatory disaster, and when considered through the connectivity of the atmosphere and in frames of global visibility, reveals the impactfulness of devastation in the skies.

The scale of devastation contributes to the framing of the *Challenger* disaster as an original accident of the engine. Evidently neither the first engine-related disaster nor the greatest in terms of casualties, the *Challenger* disaster is unique in terms of scope. President Reagan recognized the *Challenger* explosion as the first loss of life in the air for the United States' space program; and it came at a time in which the United States was still recognized as the singular global superpower, with the news of the failure circling the globe. The material condition of the event is also significant, the explosion taking place in the skies—a common, collective site.[48] There is a component of aerial catastrophes in particular that attract significant public attention. A 2016 study of aircraft disasters in the context of the digital sought to measure the impact of such events, finding the persistence of aircraft disasters in cultural memory extends for approximately forty-five years from their occurrence.[49] Numerous disasters in recent memory (such as the MH370 disappearance in 2014, the Germanwings Flight 9525 crash in 2015, and the Ukraine International Airlines Flight 752 in 2020) capture global attention in a remarkable manner, owing to their catastrophic nature, and also through connectivity across place and aerial presence. In understanding the weightedness of the *Challenger* disaster, the air presents as a crucial site for investigation, materially, and also politically and affectively. In doing so, the affective resonance of the *Challenger* disaster illuminates the identification of the disaster as an original accident of the engine, alongside presenting the potential to theorize the disaster in relation to perception, construction, and power.

Ripples of the *Challenger* Disaster

Michel Serres addresses *Challenger* in his 1987 book *Statues: The Second Book of Foundations*, where he aligns the catastrophe with the ancient Carthaginian ritual of Baal (in which children and animals would be encased in a large metal statue and burned alive). In speaking to *Challenger* in the context of his philosophical worldview, Serres states:

> When we place society on one side and science on another, we no longer see anything. A certain light, strong and focused, dazzles the eyes, whereas placing an object in light and shadow allows us to see it. Actually, we always see in this way, in the light and shadow of the real atmosphere. The pure light of the sun would burn our eyes, and we would die of cold in the darkness.[50]

While Serres is speaking here to the necessity of a holistic approach to philosophy and theory; utilizing the atmosphere as a metaphor, he locates humankind within physical and environmental atmospheric space. With these comments, he endeavors to present to his interlocutor, Latour, reasoning for his philosophical approach—one in which, through mixing, confluence, and entanglement, the necessity of thinking across category and boundary becomes apparent.[51] Serres, in his statements about science and society, and through the metaphor of the atmosphere, presents two trajectories to follow, much like contrails bisecting the sky. The first is again the recognition and subsequent transformation of the negative condition of air. Following Eddington in inverting this condition, Serres encounters the illuminatory, but not blinding, possibilities of light and shadow in the atmosphere. The second is encounter with the blinding light, burning the eyes—a trajectory that leads to and mimics certain forms of accident, catastrophe, and disaster. In the space between catastrophe and inverted condition of the atmosphere, it is possible to think through material and connectivity in apprehending the impact of aerial catastrophes, and to see into aerial space the connectivity of air itself.

Following Virilio from invention to inverse miracle, a parallel can be drawn with the understandings of English polymath Charles Babbage, who stated in 1837 his belief in "the principle of the equality of action and reaction."[52] A mathematician, considered by many to be the father of the modern computer, Babbage traced the consequences of action through the lens of the natural world, preceding contemporary physics and new materialist understandings of matter. He wrote: "Everyone who has thrown a pebble into the still waters of a sheltered pool, has seen the circles it has raised gradually expanding in size, and as uniformly diminishing in distinctness."[53] If the *Challenger* disaster is a pebble in the still waters of the sky, the circles of its catastrophe continually ripple outwards, with consequences for understandings of air and the atmosphere in the present.

The *Challenger* disaster is preserved in cultural memory, through the capture and circulation of the image of the explosion: White billowing clouds of smoke obscure the damaged shuttle, set against the backdrop of a vivid blue sky. This sky was picture-perfect and cloudless, a backdrop for the catastrophic events that occurred, but also the space in which they happened. The image itself is a capture of disaster, whether that image is in the video footage or the haunting still of white billowing explosion against the vivid blue sky. In her apprehensions of the materiality of the image, philosopher of film and new media Laura U. Marks recognizes that images are a connective tissue that bring the viewer into contact with materiality, but also that contemporary society prioritizes the ocular, and that it is its "dominance [that] parallels the dominance of idealism in western cultures."[54] This leaves air in a difficult place, visually, materially, and politically: As subject of viewing, it is absent from ocular perception until made visible. This visibility, as seen with *Challenger* and also in contemporary airscapes of ecological crisis, often arises in crisis and catastrophe.

Catastrophe and disaster act as illuminative force. In the vivid spark of the explosion and the unnatural beauty of smoke clouds, the *Challenger* disaster presents as an opportunity to tie affect to aerial disaster. Zooming out from the discrete catastrophic event, shifts in the field of disaster studies, particularly in the frame of the interpretive fiction and the necessary political components of the disaster, can be leveraged in order to consider the implications of aerial catastrophe as divorced from those initial pillars identified by Fritz.[55] An argument against the expansion of the disaster or catastrophe beyond positioning as "sudden and violent changes in the physical environment" is that temporal expansions of the category might "capture too much," leading to a lack of specificity and therefore meaning.[56] However, emergent approaches to the field of disaster studies recognize a shift in these events themselves throughout the twenty-first century, a change in their causes and conditions if viewed not simply as events but as resultant from processes.[57]

What is distinct in the twenty-first century, with reference to disasters, is a change in scope and scale as tied to globalization and to climate. The work of sociologist Eric L. Hsu in this area presents a broad overview of current debates in the field, mounting a case for temporal expansion in recognition of the complexity of disaster in the contemporary—pointing toward catastrophic climate change in particular. Importantly, questions of speed and time in the context of harm are challenged in the increasingly crucial work of environmental humanities theorist Rob Nixon. Nixon's notion of "slow violence" rethinks violence as an immediate act, toward an expansion through political, social, and material frames in order to conceptualize another, widespread, form of violence—an "accretive" violence.[58] Through readings of environmental pollution and degradation, Nixon describes this harm as "violence that occurs gradually and out of sight, a violence of delayed destruction that is dispersed across time and space, an attritional violence that is typically not viewed as violence at all."[59] This notion becomes significant when attending to aerial catastrophes, read as dispersed across borders and through cultural memory. Further, the climate crisis challenges frames of reference in the present, as throughout history it has been evident that humanity "has no experience of dealing with such combinations of scale and speed of environmental change."[60] However, while the *Challenger* explosion might be read as immediate and localized harm, the aerial nature of the event renders it dispersed through the connective tissue of air, aided by the visibility of the event read against the backdrop of a seemingly empty sky.

Understanding Aerial Space through Transit

There is a quality to aerial disasters that sees them linger in cultural memory. Further, aerial transit signifies connectivity, crossing borders, and moving between sites, traversing the skies at increasing frequency and pace. However, it is important to consider the myriad of ways in which the sky is not simply a universal commons, the ways in which airspace is controlled, regulated, and transformed. While many across the world might believe that they stare into the same sky, draw the same sky as children, breathe the same air from Earth's atmosphere, the reality is very different when mediated by structures of nationalism and control.[61] Airspace is marked, occupied, and surveilled, despite being traversed by many millions each year in passenger planes.[62]

In the work of Peter Adey, noted researcher of space, security, and mobility, understandings of the realm above Earth—the sky, the air, and the atmosphere—are constructed by political frameworks and realized through forms of militarized control.[63] This understanding incorporates the recognition of a politics of verticality, moving beyond simple cartographic representations of place as land upon a map. This politics of verticality emerges from Forensic

Architecture's Eyal Weizman, whose readings of Palestinian occupation by invading Israeli forces recognize that that place is "no longer seen as the two-dimensional surface of a single territory, but as a large three-dimensional volume."[64] Adey, with Mark Whitehead and Alison J. Williams, acknowledges the complexity of the added component of this three-dimensional volume— that of air.[65] Aerial space might be understood as diffuse and mutable, but they argue that air is mediated by grounded, rigid, and physical structures that inform not only transit and passage, but also perceptions of air itself.

Adey, Whitehead, and Williams highlight notions of governmental control as identifying aerial imagery with the view from above, produced by satellites, spacecraft, and, more commonly, drones, serving to "prioritise the view and experiences of those who occupy the elevated perspective."[66] This under-standing of the elevated perspective owes great debt to Donna Haraway's notion of "situated knowledges," which continues to have great significance in feminist and posthumanist philosophy.[67] Within the framework of situated knowledge, a methodology broadly challenging notions of neutral objectivity, Haraway presents a critique of the "god trick," "a perverse capacity ... to distance the knowing subject from everybody and everything in the interests of unfettered power." For Haraway, this distancing is an enactment of "a conquering gaze from nowhere."[68] Additionally, this gaze is one predicated upon colonial structures of power that presume and employ unfettered access to indigenous Land in the pursuit of knowledge, data, or occupation that serves this conquering gaze.[69] In interrogating the perceived neutrality of the view from above, complex structures of power and subjugation become illuminated.

Tracing the history of the "view from above," curator and historian Jeanne Haffner draws a line across time from early Roman mapping experiments to the use of balloons in the late nineteenth century, toward the recognition of connectivity and control in Henri Lefebvre's considerations of social space.[70] Within this history of the view from above, Haffner identifies a critical shift in understandings of the aerial view—as directly tied to governmental control and entangled with war (the Algerian war is presented as a crucial example of this nexus).[71] The enfolding of the aerial view with structures of coloni-zation and conflict lead Lefebvre to denounce the aerial view as an exemplar of the "hierarchical stratified morphologies" of the state.[72] Significantly, Lefebvre positions state power as "the space of catastrophe," alongside denouncements of control, surveillance, and dominance of the air.[73] Haffner recognizes the importance of Lefebvre's position in the context of the back-drop of a particularly French colonial rule of the 1960s, but emphasizes the persistence of the view from above into the present.[74] Haffner rightly recog-nizes the increasingly expanded view from above over time, acknowledging increased surveillance from the skies as tied to political power.[75]

With technological developments came the opportunity to zoom even further from the surface of the Earth, moving into the outer atmosphere and even beyond the Earth itself. In the context of the view from above, the

blinding image of the *Challenger* disaster can be read against another foundational image tethered to spaceflight. The "Blue Marble," a photograph
taken by the *Apollo 17* spacecraft in 1972 is a ubiquitous image of the Earth,
presented against the backdrop of an empty cosmos, and has come to signify
perceptions of the Earth as a whole from a distance.[76] As environmental
humanities researcher Thomas M. Lekan presents, the "Blue Marble" photograph emerged at a significant point in world history, amid ongoing conflict
in Vietnam and within a rising global environmental consciousness.[77] It
contributes to the notion of a "Whole Earth," which, within the environmental movements of the 1970s, formed part of a turn toward recognition
of global environmental connectivity and collectivization. This turn was
foregrounded by the photograph's predecessor, the "Earthrise" image taken
in 1968 by crew from the *Apollo 8* mission, in which the Earth appears as
another planet in the perspective of lunar orbit—distant, fragile, and whole.[78]
Within these understandings, the planet Earth was understood to be
"Spaceship Earth," host and home for civilization, floating amid the backdrop
of empty space.[79]

Contemporary readings of these images place them within the frame of
the god trick—distanced but imbued with power, specifically that of US
militarized imperialism. Indeed, the "Blue Marble" image itself is critiqued
as emerging from these structures, motivated by ideologies of colonization—of
conquering the frontier of space, of occupation, of a cosmic wilderness to be
explored.[80] Noted geographer Denis Cosgrave takes up a similar notion
through the conceptualization of the "Apollonian gaze."[81] In thinking through
visions of the Whole Earth, of the globe as a distanced universal subject,
Cosgrave acknowledges that the universalizing vision proposed by such
idealized views of the worlds is not neutral, but rather serves to bolster the
spread and hold of Western views, buttressing the hold of globalization in
service of the West.[82] In moving between Haraway and Cosgrave, the universalizing view from above can be read not as utopian vision, but deeply marked
by political structures of power and control, which has the potential for
destabilization in the present.

Before returning to the *Challenger* disaster, consideration of the increasing
militarization of the skies reveals the continual advancement and occupation
of the air by imperialist forces. In doing so, the air and the atmosphere is
recognized as constituted by geopolitical and technological actants. As identified, the view from above is grounded in histories of militarization, which
have become accelerated in the contemporary, supported by technological
development. Importantly, notions of surveillance from above constrict movement and behavior, tracing back to aerial bombing campaigns in World War
II.[83] Interrogations of US aerial assaults on Iraq from the early 2000s provide
an understanding of warfare tactics as tied relationships to air. Martin Coward
looks to the policy of "Shock and Awe" during this period, essentially rapid
aerial assaults targeting the Iraq military, and reads them as contributing to
"the idea of American power," and simultaneously a "targeting *from* the air

but also *of* the air—of the air-waves and airtime of global news."[84] With war being fought from above, media is littered with footage of explosions taking place on foreign shores (non-Western, othered, denigrated shores). These explosions, of aerially deployed bombs, "dazzle the eyes," and feed in to a new, grotesque sublime defined by warfare.[85] Further, modern aerial warfare is now distanced, in a similar way to the distancing of the god trick through the unmanned aerial vehicle, also known as the drone. In a confronting reflection upon the military drone, award-winning filmmaker Alex Rivera notes that the drone

> is a transnational and telepresent kill system, a disembodied destroyer
> of bodies ... The reason it has become a pop-cultural phenomenon
> and an object of fascination and study ... is that it is an incandescent
> reflection, the most extreme expression of who we are and what we've
> become generally.[86]

Reading Riviera's "we" as limited to the US citizen, these reflections highlight the drone's disembodied quality. The drone enacts a distancing from the self in the enactment of the technological reach of the state beyond borders. The drone, and the specific drone vision that emerges as a way of seeing in the contemporary, presents as a new understanding of aerial regimes of seeing.

Entangled within the complex intersections of air, technology, and geopolitics in the skies above the earth is the possibility for speculating and acting otherwise with, into, and against tools employed by regimes of surveillance and control. Specifically, recent feminist thought has sought to move beyond critiques of the god trick, toward strategic and collective interventions into technocapitalist armatures. These interventions, which take the form of open access to data, satellite live feeds, and visualizations, are read as enacting and identifying breaks with traditional dualisms that set the body apart from technology, alongside the micro/macro views that unfold in relation to surveillance from above.[87] The feminist collective open-weather (Sophie Dyer and Sasha Engelmann) is identified as one such intervention that engages with citizen science and radio technologies to capture, interpret, and represent information gleaned by structures that traditionally enforce the view from above.[88]

While open-weather tunes into existing technological structures that construct, create, and continue the view from above, a 2017 text from Gabrys regarding the Citizen Sense research project (which she leads) illuminates the potential for alternate forms of aerial registration in destabilizing existing structures that create and construct the air and atmosphere within regimes of sensing and seeing.[89] With reference to citizen science air pollution modeling and the creation of new data regarding air pollution, Gabrys reflects on the potential for "generating evidence," from which she speculates that "new forms of care could emerge through these speculative approaches for evidencing harm."[90] Such projects, emerging from feminist and environmental critiques of the intersection of data, surveillance, and ecologies, illuminate possibilities for destabilizing the view from above in the present.

The image of the *Challenger* disaster can be read retrospectively as a challenge to the view from above. While undoubtedly a project embedded within processes of capitalist, imperialist, and colonizing control, the spectacular failure of the *Challenger* space shuttle is aligned with ways of viewing it within air. The key circulating image of the disaster is a still taken from video footage captured on the ground—originally broadcast by CNN and reaching into homes across the globe. While imbued with feelings of shock and awe, and circulating in news media, the *Challenger* disaster is an oddity in how it fits within vertical, aerial, universalizing structures from above. The disaster was viewed from the ground, the backdrop read as the vivid blue skies above Cape Canaveral. This argument is not one that suggests the explosion of the space shuttle as negating US power in any way, but rather it is an image read from below, as opposed to the prioritized ascendant view. The aerial nature of this event, in the context of representations and experiences of aerial space in the context of technological and political power, is undoubtedly a key quality of the disaster.

Read from below, this disaster becomes an earthly event; read from the position of a shared sky, a shared air, it complicates projections of power through the dominance of aerial space. It fails to achieve the view from above and this failure, this catastrophe, presents as a revelatory event (to follow from Eva Horn).[91] In a rather utopian manner (one that did not anticipate the new contemporary regimes of surveillance and warfare that fill the skies), geographer David Matless recognized that aerial transit presented the opportunity for a "sky-situated knowledge" that undoes the god trick through locatedness and experience.[92] However, perhaps it is the perception of aerial disaster from below that might open up the possibility of an aerially oriented knowledge.

The *Challenger* disaster, the original accident of the engine, complicates viewing the sky, the air, and the atmosphere. As much as these sites are perceived through the lens of the aforementioned negative condition, they are cloaked in regimes of control, political subjugation, and invisible structures of surveillance. A wound in the sky—failure to ascend, launch satellites, and project US imperial control amid the Cold War—is a spark that acts as a symbol of hubris and failure.

Revelation, Reverberation, and Connectivity

Recalling the bad star of the disaster, an omen or portent, alongside Reagan's Catholic overtures in addressing the nation following the *Challenger* disaster, the tragedy of 1986 can be read through Virilio as an inverted miracle, a secular revelation—particularly in the performance of an interpretive fiction of the disaster. This revelation might be one of aerial connectivity. Through the immediate, seemingly instantaneous and defining catastrophe of the

Challenger disaster, aerial space is understood differently in light of the representations of the event. As aforementioned, Horn positions catastrophes as revelatory, and the image of the *Challenger* disaster might be revelatory too. Karen Strassler, in looking to the photographs of holy figures, notes:

> Revelatory traces, such photographs also retain the aura of their originals through a property of indexical "contagion" or "contact." Within this semiotic ideology, the indexical nature of the photographic image—its physical connection to its referent—enables it to embody and transmit the power of the photographed subject.[93]

Following Virilio's inverted miracle and secular revelations, alongside the perception of the *Challenger* as a portent, the image of the *Challenger* is read as transmission. This image might be read as the transmission of a message that reveals something of the nature of aerial catastrophes tied to the engine. Via Haraway and Cosgrave, among others, the view from above is understood as a view complicated by structures of power. Viewed from below, however, the negative condition of air becomes momentarily destabilized—no longer a backdrop for disaster, or simple negative space, but a complex site for its occurrence.

In addressing the connectivity of matter, Babbage wrote: "if the air we breathe is the never-failing historian of the sentiments we have uttered, earth, air, and ocean, are in like manner the eternal witnesses of the acts we have done."[94] The day of the *Challenger* disaster, witnesses to the event were many, and in the days following, news of the event circled the globe. In this incident of immediate visibility, in which an aerial disaster illuminated the sky momentarily, conceptions of air, as empty, negative, and forgotten space, were complicated. The air was revealed as presence, as perceptible, and as affectively significant as medium and in its own right. Further to this, spaceflight is understood as cutting through the air in a very specific way; it is a conduit to space beyond the globe, and despite the increasing politicization of space, there is something universal about space itself. As such, when a disaster such as *Challenger* occurs, it amplifies shared connectivity to the air—it ripples and reverberates in a manner that terrestrial accidents might not because there is something collective underneath the layers of the political, and now corporate, endeavors that lead to spaceflight. Reaching this understanding of a collective apprehension of air through the disaster entails considerations of how this shared connectivity is shaped by frameworks that, in turn, mediate, constrict, and control air space itself.

Throughout this chapter, air's mutability and complexity has been reinforced. Taking the exploded view, challenging the negative condition, and recognizing the possibilities enabled by speculative fictions, the boundedness of the *Challenger* disaster as discrete historical event is expanded. In doing so, a reading of the complexity of relationships to air as read from above and below (and always from inside) is undertaken. In the moment of explosion, it becomes possible to read a unitary fracture. While this might read as an oxymoron, in this case it is not. In breaking up the skies through white plumes of smoke, heat, fire, and light, the *Challenger* was an aerial view but from below. With the challenges to the notion of the Whole Earth that emerged in the context of US imperialism, and the devastation of ongoing aerial conflict, now piloted remotely from the ground, the *Challenger* is situated in a between place. The original accident of the engine revealed aerial space in the context of disaster, through its shock, awe, and visibility.

Notes

1   Bruno Latour, *From Realpolitik to Dingpolitik or How to Make Things Public* (Merve: Berlin, 2005), 10–13.
2   Latour, *Realpolitik to Dingpolitik*, 14.
3   Ten million surveyed Americans reported attending a memorial service for the *Challenger* astronauts in the weeks following the event, with a number of temporary and permanent exhibitions and memorials constructed to commemorate the loss: Jon D. Miller, "The Challenger Accident and Public Opinion," *Space Policy* 3, no. 2 (May 1987): 128–29. https://doi.org/10.1016/0265-9646(87)90009-9 Capturing the extent to which the *Challenger* disaster was perceived as a monumental global disaster was the unprecedented designation of monument sites, from the Soviet Union naming crater sites on Venus in the name of some of the astronauts, to the United States memorializing the catastrophe through the renaming of Colorado mountains. "Soviets Name Venus Craters After McAuliffe and Resnik," *Columbia Record*, February 3, 1986. A Ethel Evangeline Martin Bolden Papers, Box 7, South Carolina Library; "Peak May Honor Astronauts," *New York Times*, April 12, 1987.
4   David Shayler, *Accidents and Disasters in Manned Spaceflight* (London: Springer, 2000), 186.
5   Michael Hirsley, "Shuttle Tragedy Stuns Nation," *Chicago Tribune*, January 29, 1986, 13.
6   Neel V Patel, "Coward, "What the Challenger Explosion Means 30 Years Later," *Inverse*, January 27, 2016, https://www.inverse.com/article/10686-what-the-challenger-explosion-means-30-years-later, accessed May 5, 2023.
7   Ronald Reagan, "Reagan's Address to The Nation," History. NASA, 1986, https://history.nasa.gov/reagan12886.html.
8   Reagan, "Reagan's Address to The Nation."
9   Feliks Banel, "Challenger Disaster and Local Media Memory | Crosscut," *Crosscut*, 2011, https://crosscut.com/2011/01/challenger-disaster-local-media-memory, accessed June 25, 2022.
10 John Constable Study of Cirrus Clouds, oil painting, 1822, Victoria and Albert Museum, London.
11 "Shuttle Explosion still Mystery—Americans in Shock, Sorrow," *Honolulu Advertiser*, January 29, 1986; Michael Binyon and Christopher Thomas, "Crew Die in Shuttle Disaster—US Challenger Space Shuttle Explodes," *The Times*, January 29, 1986.
12 Nerea Calvillo, "Particular Sensibilities," *e-flux Architecture*, 2016, https://www.e-flux.com/architecture/accumulation/217054/particular-sensibilities/, accessed June 25, 2022.

13 Steven Connor, "On the Air," transcript of radio broadcast for BBC Radio, June 13, 2004, http://stevenconnor.com/onair.html, accessed June 25, 2022.

14 Connor, "On the Air."

15 Vicky Lewis, "Young Children's Painting of the Sky and the Ground," *International Journal of Behavioral Development* 13, no. 1 (1990): 49. https://doi.org/10.1177/016502549001300104 Elizabeth Coates, "'I Forgot the Sky!' Children's Stories Contained within Their Drawings," *International Journal of Early Years Education* 10, no. 1 (2002): 12. https://doi.org/10.1080/09669760220114827 José Villarroel and Xabier Villanueva, "A Study Regarding the Representation of the Sun in Young Children's Spontaneous Drawings," *Social Sciences* 6, no. 3 (August 22, 2017): 95. https://doi.org/10.3390/socsci6030095

16 Lewis, "Young Children's Painting," 50.

17 Steve Mentz, "A Poetics of Nothing: Air in the Early Modern Imagination," *postmedieval: a journal of medieval cultural studies* 4, no. 1 (2013): 38. https://doi.org/10.1057/pmed.2012.42

18 Luce Irigaray, *The Forgetting of Air in Martin Heidegger* (London: The Athlone Press, 1999), 14.

19 Irigaray, *The Forgetting of Air*, 5.

20 Mentz, "A Poetics of Nothing," 38.

21 Kenneth R. Olwig, "All That is Landscape is Melted into Air: The 'Aerography' of Ethereal Space," *Environment and Planning D: Society and Space* 29, no. 3 (2011): 521. https://doi.org/10.1068/d8409

22 Olwig, "All That is Landscape is Melted into Air," 521.

23 Arthur Stanley Eddington, *New Pathways in Science* (Cambridge: Cambridge University Press, 1935), 40.

24 Eva Horn, "Air as Medium," *Grey Room* 73 (2018): 7. https://doi.org/10.1162/grey_a_00254

25 Derek P. McCormack, *Atmospheric Things* (Durham, NC: Duke University Press, 2018), 4.

26 Gernot Böhme, "The Theory off Atmospheres and Its Applications" (translated by A.-Chr. Engels-Schwarzpaul), *Interstices: Journal Of Architecture And Related Arts* (2014): 3. https://doi.org/10.24135/ijara.v0i0.480

27 Kathleen Stewart, "Atmospheric Attunements," *Environment and Planning D: Society and Space* 29, no. 3 (June 2011): 447. https://doi.org/10.1068/d9109

28 Siobhan Carroll, *An Empire of Air and Water: Uncolonizable Space in the British Imagination, 1750–1850* (Philadelphia: University of Pennsylvania Press, 2015), 119. https://doi.org/10.9783/9780812291858

29 Michel Serres and Bruno Latour, *Conversations on Science, Culture, and Time* (Ann Arbor: University of Michigan Press, 1995), 160. https://doi.org/10.3998/mpub.9736

30 R.M. Davis, "BMJ Bans 'Accidents,'" *BMJ* 322, no. 7298 (2001): 1320–21. https://doi.org/10.1136/bmj.322.7298.1320

31 The field of disaster studies is defined as one "address[ing] the social and behavioral aspects of sudden onset collective stress situations typically referred to as mass emergencies or disasters." Michael K. Lindell, "Disaster Studies," *Current Sociology* 61, no. 5–6 (September 2013): 797. https://doi.org/10.1177/0011392113484456

32 Jacob Remes and Andy Horowitz, *Critical Disaster Studies* (Philadelphia: University of Pennsylvania Press, 2021), 2. https://doi.org/10.9783/9780812299724

33   Diane Vaughan, *The Challenger Launch Decision: Risky Technology, Culture, and Deviance at NASA* (Chicago: University of Chicago Press, 1996). https://doi.org/10.7208/chicago/9780226346960.001.0001

34   Remes and Horowitz, *Critical Disaster Studies*, 2.

35   Remes and Horowitz, *Critical Disaster Studies*, 2–3.

36   Remes and Horowitz, *Critical Disaster Studies*, 2.

37   Franz Mauelshagen, "Defining Catastrophes," in *Catastrophe and Catharsis: Perspectives on Disaster and Redemption in German Culture and Beyond*, ed. Katharina Gerstenberger and Tanja Nusser (Woodbridge: Boydell & Brewer, 2015), 172.

38   Paul Virilio and Sylvère Lotringer, *Politics of the Very Worst* (New York: Semiotext(e), 1999), 89.

39   Elizabeth DeLoughrey, "Radiation Ecologies and the Wars of Light," *MFS Modern Fiction Studies* 55, no. 3 (2009): 469. https://doi.org/10.1353/mfs.0.1630

40   Scott Gabriel Knowles and Zachary Loeb, "The Voyage of the Paragon: Disaster as Method," in *Critical Disaster Studies*, ed. Jacob A.C. Remes and Andy Horowitz (Philadelphia: University of Pennsylvania Press, 2021), 11. https://doi.org/10.2307/j.ctv1f45qvg.4

41   Neeltje van Doremalen, Trenton Bushmaker, and Dylan H. Morris, "Aerosol and Surface Stability of SARS-CoV-2 as Compared with SARS-CoV-1," Letter to the Editor, *New England Journal of Medicine* (March 17, 2020): 1564–67. https://doi.org/10.1056/NEJMc2004973

42   DeLoughrey, "Radiation Ecologies," 471.

43   Gregory Whitehead, "The Forensic Theatre: Memory Plays for the Post-mortem Condition," *Performing Arts Journal* 12 (Spring 1990): 100–01. https://doi.org/10.2307/3245556

44   William Chambers and James Donald, *Chambers's Etymological Dictionary of the English Language* (London: W. & R. Chambers, 1874), 128; François Debrix, "End Piece: Dealing with Disastrous Life," in *Biopolitical Disaster*, ed. Jennifer L. Lawrence and Sarah Marie Wiebe (London: Routledge, 2017), 258. https://doi.org/10.4324/9781315620213-21

45   Charles E. Fritz, "Disaster," in *Contemporary Social Problems*, ed. R.K. Merton and R.A. Nisbet (New York: Harcourt, 1961), 655.

46   Paul Virilio, "Surfing the Accident," interview by Andreas Ruby, *The Art of the Accident* (Rotterdam: V2_ publishing, 1998), 30.

47   Virilio, *Politics of the Very Worst*, 89.

48   Mark Everard, Ben Pontin, Tom Appleby, Chad Staddon, Enda Hayes, Jo Barnes, and J Longhurst, "Air as a Common Good," *Environmental Science & Policy* 33 (2013): 354–68. https://doi.org/10.1016/j.envsci.2012.04.008

49   Ruth Garcia-Gavilanes, Anders Mollgaard, Milena Tsvetkova and Taha Yasseri, "Memory Remains: Understanding Collective Memory in the Digital Age," *Science Advances* 3, no 4 (2016). https://doi.org/10.1126/sciadv.1602368

50   Michel Serres, *Statues: The Second Book of Foundations* (London: Bloomsbury Publishing, 2014), 148.

51   Serres and Latour, *Conversations*, 99–102.

52   Charles Babbage, *The Ninth Bridgewater Treatise: A Fragment* (Philadelphia, PA: Lea & Blanchard, 1841), 127.

53   Babbage, *The Ninth Bridgewater Treatise*, 128.

54 Laura U. Marks, *Touch* (Minneapolis: University of Minnesota Press, 2002), xvii.

55 Eric L. Hsu, "Must Disasters Be Rapidly Occurring? The Case for an Expanded Temporal Typology of Disasters." *Time & Society* 28, no. 3 (August 2019): 904–21. https://doi.org/10.1177/0961463X17701956

56 Allen H. Barton, "Disaster and collective stress," in W*hat is a Disaster? New Answers to Old Questions*, ed. Ronald R Perry and E.L. Quarantelli (Bloomington: Xlbris, 2005), 153; Enrico Quarantelli, "A Social Science Research Agenda for the Disasters of the 21st Century: Theoretical, Methodological and Empirical Issues and their Professional Implementation," in *What is a Disaster? New Answers to Old Questions*, ed. Ronald R Perry and E. L. Quarantelli (Bloomington: Xlbris, 2005), 333. https://doi.org/10.4324/9780203984833

57 Steve Matthewman, *Disasters, Risks and Revelation* (New York: Palgrave Macmillan, 2011), 5-10.

58 Rob Nixon, *Slow Violence and the Environmentalism of the Poor* (London: Harvard University Press, 2011), 2. https://doi.org/10.4159/harvard.9780674061194

59 Nixon, *Slow Violence*, 2.

60 Will Steffen, Åsa Persson, Lisa Deutsch, Jan Zalasiewicz, Mark Williams, Katherine Richardson, Carole Crumley, Paul Crutzen, Carl Folke, Line Gordon, Mario Molina, Veerabhadran Ramanathan, Johan Rockström, Marten Scheffer, Hans Joachim Schellnhuber, and Uno Svedin, "The Anthropocene: From Global Change to Planetary Stewardship," *Ambio* 40, no. 7 (November 2011): 752. https://doi.org/10.1007/s13280-011-0185-x

61 See Peter Sloterdijk, *Terror from the Air* (Cambridge, MA, and London: The MIT Press, 2009).

62 The Australian Refugee Council, for example, notes that despite the United Nations 1954 Convention regarding the rights of stateless people declaring stateless people the right to travel, the reality is very different, with many unable to access this right. Additionally, the Council notes that estimates of stateless people globally is in excess of 10 million. See "Statelessness," *Refugee Council of Australia*, 2022, https://www.refugeecouncil.org.au/stateless/; Peter Adey, *Air: Nature and Culture* (London: Reaktion Books, 2014); Mark Whitehead, *State, Science & the Skyes* (Oxford: Wiley-Blackwell, 2009); Timothy Choy, "Air's Substantiations," in *Lively Capital: Biothechnologies, Ethics and Governance in Global Markets*, ed. Kaushik Sunder Rajan (Durham, NC: Duke University Press, 2010).

63 Peter Adey, *Aerial Life: Spaces, Mobilities, Affects* (Chichester: Wiley-Blackwell, 2012), 85.

64 Eyal Weizman, "The Politics of Verticality," *Mute* (2004), https://www.metamute.org/editorial/articles/politics-verticality.

65 Peter Adey, Mark Whitehead, and Alison J. Williams, "Introduction," in *From Above: War, Violence and Verticality* (London: C. Hurst & Co., 2013), 2. https://doi.org/10.1093/acprof:oso/9780199334797.001.0001

66 Adey, Whitehead, and Alison, "Introduction," 11.

67 Donna Haraway, "Situated Knowledges: The Science Question in Feminism and the Privilege of Partial Perspective," *Feminist Studies* 14, no. 3 (1988): 575–99. https://doi.org/10.1093/acprof:oso/9780199334797.001.0001

68 Haraway, "Situated Knowledges," 581.

69 Max Liboiron, *Pollution is Colonialism* (Durham, NC: Duke University Press, 2021), 1–38. https://doi.org/10.1515/9781478021445

70   Jeanne Haffner, *The View from Above: The Science of Social Space* (Cambridge, MA: The MIT Press, 2013), 108. https://doi.org/10.7551/mitpress/7878.001.0001

71   Haffner, *The View from Above*, 109.

72   Henri Lefebvre, "The State in the Modern World (1975)," in *State, Space, World: Selected Essays*, ed. Neil Brenner and Stuart Elden (Minneapolis: University of Minnesota Press, 2009), 118.

73   Lefebvre, "The State in the Modern World," 118.

74   Haffner, *The View from Above*, 113–14.

75   Lefebvre, drawing from Karl Marx, also recognizes that "There is a religiosity of State linked to the existence of the State itself because the State is, in relation to real life, in the same relation as the sky to the earth, which is to say above real life, looming over it." This is simultaneously a metaphorical position and a reflection of conditions that becomes increasingly true in the contemporary period. Henri Lefebvre, "The Withering Away of the State: The Sources of Marxist-Leninist State Theory (1964)," in *State, Space, World: Selected Essays*, ed. Neil Brenner and Stuart Elden (Minneapolis: University of Minnesota Press, 2009), 75.

76   Haraway, "Situated Knowledges," 581.

77   Thomas M. Lekan, "Fractal Eaarth: Visualizing the Global Environment in the Anthropocene," *Environmental Humanities* 5, no. 1 (2024): 175–76. https://doi.org/10.1215/22011919-3615469

78   Life Magazine, *100 Photographs that Changed the World* (New York: Time Inc., 2003), 172.

79   See Kenneth E. Boulding et al., *Human Values on the Spaceship Earth* (New York: Council Press for Commission on Church & Economic Life, National Council of Churches, 1966).

80   Stefan Helmreich, "From Spaceship Earth to Google Ocean: Planetary Icons, Indexes, and Infrastructures," *Social Research* 78, no. 4 (2011): 1214. https://doi.org/10.1353/sor.2011.0042

81   Denis Cosgrove, *Apollo's Eye: A Cartographic Genealogy of the Earth in the Western Imagination* (Baltimore, MD: Johns Hopkins University Press, 2003), ix-xiii.

82   Denis Cosgrove, "Contested Global Visions: One-World, Whole-Earth, and the Apollo Space Photographs," *Annals of the Association of American Geographers* 84, no. 2 (1994): 270–94. https://doi.org/10.1111/j.1467-8306.1994.tb01738.x

83   Torben Möbius, "World War II Aerial Bombings of Germany: Fear as Subject of National Socialist Governmental Practices," *Storicamente*, 11 (2015): 21. https://doi.org/10.12977/stor606

84   Martin Coward, "Networks, Nodes and De-Territorialised Battlespace: The Scopic Regime of Rapid Dominance," in *From Above: War, Violence and Verticality*, ed. Peter Adey, Mark Whitehead, and Alison J. Williams (London: C. Hurst & Co., 2013), 97. https://doi.org/10.1093/acprof:oso/9780199334797.003.0005

85   Coward, "Networks, Nodes and De-Territorialised Battlespace," 97.

86   Malcolm Harris, "Border Control," *New Inquiry*, 2022, https://thenewinquiry.com/border-control/, accessed May 3, 2023.

87   Sasha Engelmann, Sophie Dyer, Lizzie Malcolm, and Daniel Powers, "Open-Weather: Speculative-Feminist Propositions for Planetary Images in an Era of Climate Crisis," *Geoforum* 137 (2022): 239. https://doi.org/10.1016/j.geoforum.2022.09.004

88   Engelmann, "Open-Weather," 237–47.

89 Jennifer Gabrys, "Citizen Sensing, Air Pollution and Fracking: From 'Caring about Your Air' to Speculative Practices of Evidencing Harm," *Sociological Review* 65, no. 2 (2017): 172–92. https://doi.org/10.1177/0081176917710421

90 Gabrys, "Citizen Sensing," 186.

91 Eva Horn, "The Last Man: The Birth of Modern Apocalypse in Jean Paul, John Martin, and Lord Byron," in *Catastrophes: A History and Theory of an Operative Concept*, ed. Nitzan Lebovic and Andreas Killen (Berlin and Boston, MA: De Gruyter Oldenbourg, 2014), 55. https://doi.org/10.1515/9783110312584.55 Eva Horn, *The Future as Catastrophe: Imagining Disaster in the Modern Age* (New York: Columbia University Press, 2018), 12. https://doi.org/10.7312/horn18862

92 David Matless, "The Uses of Cartographic Literacy: Mapping, Survey and Citizenship in Twentieth-Century Britain," in *Mappings*, ed. Denis Cosgrove (London: Reaktion, 1999), 212.

93 Karen Strassler, *Refracted Visions: Popular Photography and National Modernity in Java* (Durham, NC: Duke University Press, 2010), 282. https://doi.org/10.2307/j.ctv11cw7gs

94 Babbage, *The Ninth Bridgewater Treatise*, 127.

PART II

AIR AND ART
IN TIMES OF CRISIS

4

Speculative Fiction, Atmotechnic Ecology, and the Afterlife of Romantic Air

Siobhan Carroll

> By using violence against the very air that groups breathe, the human
> being's immediate atmospheric envelope is transformed into something
> whose intactness or non-intactness is henceforth a question.
>
> Peter Sloterdijk, *Terror from the Air*[1]

Introduction: The Transformation of Air

In *Terror from the Air* (2009), German philosopher Peter Sloterdijk portrays
the twentieth century as an era distinguished by a new relationship to air.
Following the deployment of gas as a weapon during World War I, he argues,
"the human being's immediate atmospheric envelope" was transformed from
a neutral environment "into something whose intactness or non-intactness
is henceforth a question."[2] This new atmospheric vulnerability led to "air and
atmosphere" becoming the objects of "explicit consideration and monitoring"
in fields ranging from medicine to cultural theory, and, he claims, to atmotechnics (technological and architectural control over air) becoming an
important tool of the state.[3] Sloterdijk's insights into the politics of air are
strikingly relevant to our current historical moment, in which we are still
feeling the effects of the dramatic global reordering caused by an airborne
pandemic. Prior to 2020, many of us may have felt secure in what Gregers
Andersen, following Sloterdijk, calls "safe" air; lulled by familiar practices of
consumption into forgetting "that air-conditioning and the technological
tempering of climate provide us with protection and comfort in an environment we might otherwise experience as hostile."[4] The 2020 pandemic
disrupted our relationship with air, challenging the ability of nation-states,

Siobhan Carroll, "Speculative Fiction, Atmotechnic Ecology, and the Afterlife of Romantic Air" in: *Imagining
Air: Cultural Axiology and the Politics of Invisibility.* University of Exeter Press (2023). © Siobhan Carroll.
DOI: 10.47788/ZJTA2307

corporations, and disciplines to "control" air and therefore their ability to protect the constitutive members of their communities. By the same measure, it took away something that many people, without realizing it, had seen as their right: the ability to access and circulate in "safe" air; an ability that in many cases relied on one's participation in an economic system (e.g., one must pay to air condition one's home) but not on measures that affected one's sense of bodily freedom (e.g., masking).[5] Viewed through the lens of Sloterdijk's theories, many of the conflicts that we have seen in recent years would thus appear inflected by the pandemic's upending of atmotechnic life.

While Sloterdijk may be essentially correct in identifying atmotechnics as a central feature of late capitalism, however, the history he constructs tracing this development back to World War I is clearly problematic. As Jesse Oak Taylor has argued, we have been identifying modernity with the human capacity to create artificial climates since at least the nineteenth century, albeit as a climate most easily seen through the specter of smoke pollution.[6] Moreover, as my scholarship on Romantic hot air balloons indicates, the atmosphere was a locus of national insecurity long before the gas attacks of Ypres. Sloterdijk's perception that the atmosphere only became "visible" in the twentieth century is thus a product of a historical oversight that imagines the contested atmosphere as a relatively recent invention. Sloterdijk's oversight should remind us of the similar oversight of ecocritics who assume climate change anxiety to be an exclusively twenty-first-century preoccupation; a perception aided, as Amitav Ghosh has suggested, by a myopic focus on realist texts as the primary venue for articulations of such anxieties. As Ghosh observes regarding the limitations of the mainstream novel, other genres of writing, namely "science fiction,"[7] are quite comfortable representing climate change, albeit at a remove from daily life. I would argue more broadly that speculative fiction—the umbrella form that encompasses science fiction—has a long history of making "invisible" atmospheric threats such as plague and climate change distinctly visible. If over the next few years, literary critics want to observe cultural shifts occurring in pre- versus post-pandemic representations of air, we should make sure to attend to speculative fiction's expressions of our culture's atmospheric imaginary.

In this chapter, I focus on speculative fiction as a particularly important archive in the cultural history of air, one that registers significant shifts in the way we think about the spatiality and nature of transparent air. The term "speculative fiction" has gained popularity in recent decades as a useful way of grouping together nonrealist genres such as science fiction, fantasy, and the weird, which share a common commitment to "imagining things otherwise than they are"[8]. Popularized in the 1940s by Robert Heinlein,[9] an influential science fiction writer and an aeronautical engineer, speculative fiction has long shared a certain association with the atmosphere, being used intermittently in the first part of the twentieth century to describe the extrapolative futurist fictions that imagined flight.[10] Derived from the Latin participial stem of *specular*, "to spy out, watch, examine, observe,"[11] and closely related to the Latin word for "watchtower," "speculative" invites us to think of the form as envisioning things that are difficult to see: magical or future worlds, or, I would argue, transparent air.

This chapter traces the evolution of speculative fiction's representation of transparent air from the Romantic period through to our own moment of pandemic crisis. I begin with Mary Shelley's *The Last Man* (1826), a Romantic speculative novel that "makes visible" the entanglements of transparent air in its portrait of an airborne plague. Shelley's novel serves as an early example of Anthropocene fiction, inviting its readers not only to attend to transparent air, but also to question whether human action can affect the planetary atmosphere. I then consider a series of significant speculative works that followed *The Last Man*, demonstrating that works such as M.R. James's "Oh, Whistle, and I'll Come to You, My Lad" (1904) revise Romantic understandings of humans' ecological entanglements to address new developments in the atmotechnics of national airspace. In the third part of the chapter, I analyze the atmotechnic politics of recent science fiction novels such as *Station Eleven* and *A Song for a New Day*, arguing that they showcase a new understanding of atmospheric insecurity, one that articulates a conception of what, following Sloterdijk, we might call atmotechnic citizenship. I conclude with a reading of the altered conditions affecting certain pandemic-era speculative works, which dramatize both the ongoing legacy of Romantic air and the pandemic's impact on these attempts to represent aerial entanglements. By looking at these and similar speculative works, I suggest, we can trace our evolving conceptions of ecological responsibility, aerial connectivity, and atmospheric space into the twenty-first century, and into a new era of atmospheric anxiety.

The Last Man and Romantic Aerial Speculation

Depicting events transpiring in the later twenty-first century, Mary Shelley's *The Last Man* (1826) can be read as an early work of Anthropocene science fiction that probes the relationship between the human race and planetary-scale Nature via its representation of transparent air. Confident in the progress of human society, characters in the novel envision human and natural energies working in harmony to alter the climate of the world and produce a "paradise" on earth. From the perspective of the early nineteenth century, these characters do indeed wield a surprising amount of technological power over Nature, which is manifested in their seeming domination of the atmosphere. This is most obviously manifested in characters' use of sophisticated hot air balloons, which traverse the globe, their "feathered vans cleaving the unopposing atmosphere."[12] Their seeming control over a cooperative natural world contributes to these future Britons' hubris, encouraging Raymond, the Byronic would-be liberator of Constantinople, to invade the city despite his troops' concern over its "pestilential air."[13] The subsequent battle results in an explosion that propels the dust of Constantinople into the atmosphere, making formerly invisible air the object of perception and debate. Thereafter, the human race begins to succumb to a mysterious plague that kills everyone it touches. Even as the plague spreads across the globe, however, Britons remain confident that the climate of their island will not be touched by its ravages. Interrupting a debate over climate improvement, one cynical Briton observes that "it is as wise to discuss the probability of a visitation of the plague in our well-governed metropolis, as to calculate the centuries which must escape before we can grow pine-apples here in the open air."[14] However, his astute words are ignored by his companions. Despite the fact that their hot air balloons testify to the atmospheric connection between their island and the other parts of the globe, and despite the fact that they believe in the permeability of the human body to the atmosphere, Mary Shelley's future Britons continue to imagine their island climate as immune to the effects of this airborne plague. Soon, however, the atmosphere turns against them. Strong winds move forcefully over the world, such that "the frail balloon dares no longer sail on the agitated air;"[15] and the plague is distributed still further. Humans "call ourselves lords of the creation ... and we allege in excuse of this arrogance, that though the individual is destroyed, man continues for ever,"[16] the narrator reflects. Now, however, Nature has turned against humanity, which is annihilated by an "air ... subject to infection,"[17] circulating in a decidedly planetary atmosphere. As the title of the novel suggests, the human race is reduced to a single survivor who records the death of the human race, and the triumph of Nature over her would-be masters.

Shelley's novel reflects several historical shifts in Britons' understanding of the atmosphere. Whereas in the eighteenth century, climate was typically thought of as a local, or at most national, set of characteristic weather patterns, Shelley's description of both of the human race's possible ability to alter the climate and the potential effect of a local event on the *planet's* atmosphere displays a new understanding of the atmosphere as global, and potentially subject to the effects of human action. As the presence of hot air balloons indicate, this shift in geo-imaginary representation had in part been instigated by the invention of balloon flight, which at a stroke made the transparent atmosphere above Britons' heads visible as a site of human movement and therefore as a site vulnerable to military invasion. Likewise, the traumatic impact of 1816's mysteriously disturbed atmosphere is visible both in Shelley's descriptions of the turbulent atmosphere and in the questions she implicitly raises about the possible effects of human improvements on the air (questions that were raised more directly about the 1816 climate catastrophe).[18] In *The Last Man*, the causal links between Raymond's actions and the plague are never conclusively determined. Can human action alter the climate of the globe, as *The Last Man's* idealistic characters suggest? Is human action to blame for the turbulent atmosphere that threatens the survival of the human race? True to the speculative nature of its future-set narrative, the novel invites the Romantic reader to wonder about the possible connections between human agency and environmental response.

While Sloterdijk argues for atmotechnic society as a distinctly twentieth-century invention, Mary Shelley's novel is clearly exploring, and exploding, nascent fantasies of aerial control. While characters debate the prospect of growing pineapples in England, the more mundane features of their landscape display their growing control over their environment. In addition to the showy technology of hot air balloons, our attention is drawn to descriptions of domestic climate control, such as the "heated room" that protects characters from the elements.[19] It is understandable then, that these far-future characters imagine themselves as having triumphed over the human body's fundamental vulnerability to the elements. The triumph of the plague and the destruction of this illusion of aerial immunity might thus be read as highlighting the fragility of atmotechnic identity even as (thanks in part to technologies such as hot air balloons) human aerial control becomes imaginable. As the nineteenth century progressed, speculative fictions ranging from Jules Verne's *Five Weeks in a Balloon* (1863) to M.P. Shiel's "last man" novel *The Purple Cloud* (1901) would continue to imagine the spectacle of a world newly accessible to balloon and, later, dirigible flight.[20] It is not until the early decades of the twentieth century, however, that we see a series of fictive responses to the invention of airplanes that signal yet another shift in the atmospheric and environmental imaginary.

Early Twentieth-Century Speculative Fiction and the Return of Romantic Air

In 1903, the Wright Brothers' demonstration of sustained flight launched a new era of atmospheric insecurity in speculative fiction. In trying to envision the implications of this new technology, some of the most famous Victorian-era speculative fiction authors understandably looked to the nineteenth century for inspiration, drawing on Romantic-era tropes from the age of the first air balloons to imagine the possibilities of twentieth-century airspace. In H.G. Wells's *The War in the Air* (1908), for example, a wave of balloon ascents over London signal a new era in which airplane flight, and therefore warfare, are imminent. Harking back to the first era of British balloon flight, Wells depicts the invention of the first effective airplane as happening in the shadow of the Crystal Palace.[21] Similarly, Arthur Conan Doyle's "The Horror of the Heights" (1913) tries to imagine a new aerial era by using geo-imaginary tropes from the early nineteenth century. "Dressed like an Arctic explorer,"[22] Doyle's aeronautical explorer discovers a new atmospheric region above 40,000 ft, an atopic environment populated by aerial jellyfish and other creatures that resemble maritime lifeforms. Doyle's story clearly positions the upper atmosphere as a mysterious space that, like the ocean, is increasingly open to imperial adventure. These stories clearly engage with nascent concept of twentieth-century airspace in ways that call back to Romantic treatments of the atmosphere as a novel form of geographical space.

Yet while these fictions represent speculations about airspace, they do not introduce the vulnerable atmotechnic individuals about whom Sloterdijk theorizes, the inhabitants of modernity who fear "the idea that a small breakdown would lead to suffocation."[23] *The War in the Air* is interested in oxygen only as a component in explosive bullets, while "The Horror of the Heights" actively refuses to imagine the body of its heroic protagonist as being dependent on air. Indeed, on entering the upper atmosphere, Doyle's explorer finds that he doesn't need the oxygen he has brought with him, for "even without my oxygen inhaler I could breathe without undue distress."[24] Both these stories imagine the atmosphere as a space for imperial heroism and potential international warfare, but not as the location of air-as-material, with its own agentic effects on the human body.

To look for early twentieth-century examples of atmotechnic vulnerability, we should instead look to the atmospheric ghost stories by authors such as M.R. James and Algernon Blackwood, who laid the groundwork for the speculative subgenre later known as the Weird.[25] Blackwood's "The Willows" (1907) and "The Wendigo" (1910) both feature travelers discovering their vulnerability to strange supernatural forces identified with a foreign wind.[26] However, it is M.R. James's classic "Oh, Whistle, and I'll Come to You, My Lad" (1904)—a much-reprinted ghost story cited by H.P. Lovecraft as modeling an important strand of supernatural fiction—that most compellingly explores this premise in relation to the international atmosphere. Unlike Blackwood's stories, which take place abroad, James's tale of aerial danger takes place in England, and suggests the artificiality of claims to defend the British citizen from foreign environments.

James's "Oh, Whistle, and I'll Come to You, My Lad" is an unsettling story of aerial disturbance that invokes the legacies of older modes of aerial thinking but moves decidedly beyond them. The title of James's story is taken from a poem (a "Scottish air") by Romantic writer Robert Burns, and thus might itself be said to slyly suggest the ongoing relevance of the "air" of ages past to the modern British citizen. More interesting than the idea of the return of old air, however, is the horror the story invokes via its depiction of the invasive effects of "foreign" air on the body of a British citizen in domestic space. It is this imagery that allows us to read James's tale as one of the first articulations of atmotechnic anxiety in twentieth-century speculative fiction.

A brief summary of James's story will suggest some of its aerial and environmental concerns. Professor Parkins, the story's protagonist, travels to the seaside town of Burnstow for his end-of-term break. While there, he idly investigates some ruins on the beach that are rumored to have belonged to the Knights Templar. Digging into the masonry, he discovers a dark hole from which he withdraws a metal whistle. He carries his find back to his room at the Globe Inn, noting as he does so that a mysterious figure appears to be walking behind him on the beach. Examining the whistle in his room, he blows it twice. The violent winds that follow rattle the inn, disturbing not just the professor but the other guests, who, the next day, discuss the old superstition concerning magic-users who can "whistle for the wind"—a superstition that, the professor argues, arose in the absence of modern meteorology. Why, he recalls, just the night before, "I blew a whistle twice, and the wind seemed to come absolutely in answer to my call."[27] Parkins is keen to dismiss the idea that he might possibly have agency over the atmosphere surrounding him, but his fellow guests are less certain. After several other mysterious incidents, the story culminates in a famous scene in which Parkins, having lain down to sleep, is horrified to see a figure rise from the empty bed beside him. A creature formed out of bedlinen hunts Parkins across the room, and nearly forces him out the window to his death. Parkins is saved by the timely entrance of the Colonel, a fellow guest at the inn, who ends the haunting by throwing the offending whistle into the sea. In the denouement, we are assured that the creature that so horrified Parkins appears to have been composed entirely of air and was therefore incapable of doing real harm. "There seemed to be absolutely nothing material about it," the characters reflect afterward, "save the bed-clothes of which it had made itself a body."[28] This reassurance rings hollow for Parkins, who remains haunted by this encounter for some time afterward.

Read in the light of twenty-first-century concerns regarding the Anthropocene and atmotechnic dwelling, James's story would appear to turn on the realization that the human cannot in fact be abstracted from the nonhuman environment. Parkins, we are told, is a "Professor of Ontography,"[29] a geographer who studies how living creatures react to their physiographic environments, but for the majority of the story he appears stubbornly oblivious to his own dangerous enmeshment in his local ecology. Part of Parkins's resistance to recognizing his own peril comes from his investment in scientific objectivity, which, in addition to prompting him to discount superstitious explanations for his observations, also leads him to be invested in viewing the scientific observer (himself) as though he were a disembodied being separate from the material conditions he is observing.[30] The story indicates the falsity of this view, not only in its climax, but also in the repeated descriptions that signal the impossibility of separating "human" spaces from natural ones. Even before Parkins is pursued by a monstrous figure of air, we are provided with examples of the natural environment intruding on quasi-domestic space: the sea eroding the shore, coming ever nearer to the space of the Globe Inn; the unwelcome moonlight that streams into Parkins's bedroom and that he seeks to block out. The aerial creature's intrusion into quasi-domestic space is therefore just one of several intrusions of the natural into a space designated as belonging solely to the human.

Parkins's denial of his embodied enmeshment in the environment leads him to also deny his role in causing unwanted environmental effects; a denial that I suggest we can read in light of Anthropocene anxieties over the impact of humans on complex ecological systems. The scene in which he digs out the whistle from its buried dwelling place can be read as a scene of fossil-fuel extraction, in which an ancient source of power is permanently removed from the ground, with destructive results.³¹ However, Parkins subsequently refuses to entertain the idea that there is a relationship between his use of an object that he has extracted from the Earth and the subsequent appearance of mysterious aerial phenomena. He refuses, furthermore, to admit that there is a relationship between his body and the one that trails him on the shore; refuses to admit that his whistle-blowing may have raised the wind clawing at the inn; refuses, in short, to admit that he may be responsible for initiating a change in his environment that is now returning to haunt him. These plot points help set James's story apart from other Weird tales from the same era. They help make it into what I would call an early example of Anthropocene horror—a story in which the protagonist's denial of human ecological agency becomes one of the main engines of the narrative's ratcheting tension. Whereas in Mary Shelley's *The Last Man* the reader must puzzle along with the protagonist regarding the possible connection between human action and environmental effects, in James's "Oh, Whistle, and I'll Come to You, My Lad" the reader grows increasingly alarmed at the distinctly twentieth-century protagonist's refusal to confront his ecological responsibilities as manifested in air.

Where James's story most aligns with the Romantic air depicted in Shelley's novel is in the story's insistence on the vulnerability of the human body to its local atmosphere. As Jan Golinski observes in his history of British weather culture, many believed in the eighteenth century that everything from one's libido to one's mental health could be affected by the weather and by local changes in temperature and air-pressure.³² In comparison, even in our pandemic moment, most inhabitants of the twenty-first century are eager to see their bodies as impermeable to the effects of local air. Thus, Parkins's stubborn insistence on seeing himself as separate from the environment reflects not only a problematic ecological stance, but also a distinctly twentieth-century view of the relationship between the human body and air—a view that the narrative is eager to trouble. It is striking, for example, that the terrifying specter at the heart of James's story is made of air, and that many of its actions, far from being overtly horrifying, appear aimed at making us notice the similarity between its aerial body and Parkins's own. The creature appears to adopt a human form almost as soon as Parkins claims the whistle, but rather than using this body to attack him, it spends most of the story mirroring Parkins's actions to uncanny effect: It walks along the beach behind him; it sleeps beside him on the twin bed that lies parallel to his own; it is mistaken for Parkins by the maid who changes the linen. Part of what is horrifying about this *doppelgänger* may be that its mirrorings suggest the degree to which Parkins's body is *also* constituted by air. The human body that pretends to be abstracted from the aerial environment is in fact dependent on it and at the mercy of its exchanges, which Parkins can impact but cannot ultimately control.

There also seems to be something very new, and distinctly twentieth century, about the shock expressed in the story regarding the intrusion of air into the quasi-domestic space of Parkins's bedroom. Parkins is staying in an old inn whose name, the "Globe,"[33] calls attention to the ways in which national domestic space has always one of international circulation. And yet Parkins appears to expect his inn room to be sealed off from the rest of the world: As a paying customer, he expects his room to exist in a bubble immune to the intrusions of servants (he is disturbed at the thought that servants might enter his room when he is not present) and also to the intrusions of the external atmosphere. Read through the lens of Peter Sloterdijk's arguments in other words, Parkins appears to believe himself entitled to occupy a climate-controlled atmotechnic space protected from the external environment—a belief that the aerial entity permanently disrupts. Insofar as it depicts the failure of the seemingly intact nation to protect the middle-class individual from an atmospheric incursion accidentally summoned through his own extractive practices, M.R. James's "Whistle and I'll Come to You, My Lad" thus reads as an early meditation on the false promises of atmotechnic citizenship. In the age of the Anthropocene, neither the nation nor the global system can successfully protect its middle-class citizens from an altered and dangerous atmosphere inflected by human action.

Pre-Pandemic Atmotechnic Speculative Fiction

In arguing for air-controlled spaces as the quintessential site of global capitalism, Sloterdijk has identified the "space station" as a "key metaphor for the social architecture of the coming world age." We might therefore expect science fiction, the genre that typically depicts space stations, to play an outsize role in the fictional representation of atmotechnic spaces. Indeed, late twentieth-century and twenty-first-century science fiction media is full of examples of moments at which the reader, viewer, or player is asked to notice transparent air. Narratives about space travel are prone to dramatizing the vulnerabilities of atmotechnic societies, as in Andy Weir's *The Martian*, where substantial portions of the narrative are dedicated to the protagonist's attempts to regulate the different gases in his environment, or in movies such as *Gravity* or *Interstellar*, which create narrative drama from the prospect that a character might run out of air.[34] However, transparent air is also featured outside space-set science fiction. The popular fantasy TV series *Avatar: The Last Airbender* imagines a protagonist able to wield transparent air as a weapon as well as a means of conveyance;[35] eco-horror films such as *The Happening* dramatize the dangers hidden in seemingly transparent air;[36] and near-future science fiction films such as *Contagion* were so vivid in their depictions of airborne virus circulation that they served as cultural touchstones during the early days of the COVID-19 pandemic.[37] While different genres of speculative fiction have addressed air circulation differently, their nonrealist mechanisms have consistently allowed for the unexpected highlighting of air as a transparent medium.

In considering where speculative fiction was right before the 2020 pandemic, I turn to two prize-winning near-science fiction novels from the last decade, both of which are concerned with the function of artforms under the conditions of twenty-first-century global capital and both of which process these concerns by depicting the impacts of a global airborne pandemic. Emily St. John Mandel's *Station Eleven* (2014) meditates on the pursuit of art before and after the world is devastated by a new flu virus,[38] while Sarah Pinsker's *A Song for a New Day* (2019) focuses on the performance of music in a world irrevocably changed by an airborne virus and the subsequent restrictions imposed by a right-wing government eager to suppress discontent.[39] However, these two novels take radically different stances regarding the relationship of community to globalized and digital media, a stance that plays out in their different representations of air.

In Mandel's *Station Eleven*, the atmosphere is identified as a site of cosmopolitan exchange, invoked in conjunction with globalization, digital media, and other forms of connected life. The novel's preoccupation with air emerges in its repeated dating of events to before and after "the end of commercial air travel,"[40] in its depiction of an airport as the most stable civilization center to emerge in the post-apocalypse, and in its repeated mourning of a global, digital culture mediated by air. Post-pandemic, one character tries to explain the internet to a younger survivor by likening it to information drifting "all around you ... through the air like pollen on a summer breeze."[41] In a world devastated by the impact of an airborne virus, survivors look back nostalgically on the world of digital virality that circulated in an international atmosphere to which they no longer have access.

While the novel nods to the potential alienations felt by atmotechnic inhabitants of late-capitalism, it also argues for globalized atmotechnic spaces as the site both of dwelling and artistic production. Miranda, the artist whose graphic novel gives *Station Eleven* its title, is inspired not by traditional artistic spaces, like her boyfriend's art-strewn apartment, but by the atmo-technic spaces of corporate offices and airplanes. Miranda's comic book is itself a science fiction story about atmotechnic spaces with limited air, in which characters muddle through their relationships in a world devastated by ecological disaster. These repeated elegies for atmotechnic society are woven together in an implicit argument for the compatibility of artistic production with the twenty-first-century digital economy. Rather than positing global capitalism as the natural enemy of authentic artistic produc-tion, *Station Eleven* imagines future scenes of mourning globalization in order to argue for art's compatibility with late capitalist culture, and to maintain art's necessity as the locus for identity in a globalized capitalist world.

A Song for a New Day is similarly concerned with the status of art in a globalized digital culture mediated by atmotechnic spaces, but it is far more ambivalent, if not critical, of the interactions between art and globalized capitalism. Air, in this novel, is a fundamental unit of human community, which viruses and capitalism corrupt and which digital media cannot replace. We primarily encounter atmotechnic space through the character of Rosemary Laws, a young woman who, more than a decade after the pandemic, can barely remember the world as it was beforehand. A Zoomer *avant la lettre*, she struggles to manage her anxiety when exploring the world of unlawful public gatherings in Baltimore. Attending an unlicensed concert, she panics over the phenomenon of shared air, wondering how her fellow concertgoers can "stand being in the same room" where they cannot help but experience "the air they displaced, somebody's cologne, somebody else's sweat."[42] Seeing even small groups of strangers together, she cannot help but wonder at their confidence in mingling when they don't know "if any of them had some new superbug, if a single sneeze might endanger the whole room."[43] However, Rosemary's love of music leads her to recognize that there is an incalculable benefit to the community formed by live performance, one that, now that the pandemic has passed, outweighs the security offered by digitally mediated experiences. At one live concert the guitars are described as swallowing "every inch of space in the room, filling the air, replacing the oxygen in her lungs."[44] Rosemary acutely feels her loss of "safe" air, but music metaphorically takes its place, offering her an alternative means of sustenance, an alternative medium of community, in post-pandemic times.

The "unsafe" but life-sustaining air of the live music scene contrasts with the "safe" air of the atmotechnic spaces to which Rosemary, secretly working as a corporate music scout, has access. While she is initially impressed by her air-conditioned hotel room, she instinctively lies about her accommodations when speaking to members of the underground music scene, knowing that her access to a globalized, atmotechnic space will mark her as an outsider. As Rosemary gradually ventures beyond her ideological confines, she begins to imagine herself as possessing her own atmotechnic space in crowds—an "invisible bubble" that surrounds her and keeps her safe from the air of others.[45] This mental trick for controlling Rosemary's anxiety also functions in the text as a marker of her growing independence from her employer: Perhaps she is not as dependent on global corporations for her sustenance as she originally thought. When she realizes that her corporation plans to destroy the live music scene to eliminate competition—a strategy executed by using pandemic-era laws aimed at safeguarding air—Rosemary refuses to continue in her role as a passive enabler of corporate hegemony. Her ethical journey is inspired in part by Luce, a musician and organizer of the live music community that Rosemary's actions have disrupted. Committed to creating and maintaining community outside corporate and state spaces, Luce treats air as a fundamental unit of community. In one striking passage in the book, she argues that what music and presumably other forms of art do is occupy air, connecting people through something other than viral contagion or fearmongering politics. Music, she argues, is a "virus and a vaccine and a cure,"[46] a rival to, and an inoculation against, the fascist politics that appear to dominate the dystopian United States that she and Rosemary occupy. In *A Song for a New Day*, air thus becomes an object of contest: Powerful global corporations and their government allies argue for their ability to provide atmotechnic security, while musicians and their local communities fight to maintain their ability to access unsafe un-atmotechnic spaces in which music and people can circulate freely. Like *Station Eleven*, *A Song for a New Day* identifies atmotechnic spaces with global capitalism, but the novel's investment in local community leads it to view those spaces with suspicion. This political difference no doubt contributed to the differing fates of pandemic-era *Station Eleven* and *A Song for a New Day* television adaptations: Whereas HBO released its *Station Eleven* series to mixed reviews, the proposed adaptation of *A Song for a New Day* appears to have stalled, in part because of concerns over its depiction of the politics of an airborne pandemic.[47]

Conclusion: Art in the Time of Coronavirus

I came to this project in 2021, during a visit to the PHI art museum in Montreal. An advertisement for a virtual reality (VR) artwork had caught my eye, as it seemed to draw together many of my interests—Romantic air, speculative fiction, and ecology—in a twenty-first-century installation. The installation in question was Marshmallow Laser Feast's *We Live in an Ocean of Air*, a work that, as its title suggests, seeks to make its participants reflect on the "invisible—but fundamental—connection" provided by air, which brings "animals and plants, the human and natural worlds" together (see Fig. 4.1).[48] Set in a pixelated version of California's Sequoia National Park, this VR experience requires participants to dress in heavy VR equipment that resembles the scuba gear needed for underwater exploration. As the visitor inhales and exhales, the VR screen in their "diving mask" represents their breath as clouds of particles, which flow around the exhibition space and are taken up by the material (human) and virtual (tree) bodies in the

Fig. 4.1. Marshmallow Laser Feast's *We Live in an Ocean of Air* uses VR to represent visitors' aerial entanglement with the environment. Image used with permission of Marshmallow Laser Feast, 2022.

room. Visitors are encouraged to interact with the air, using their hands to trace their breath as it flows into the enormous virtual redwoods that tower around them. By cultivating visitors' virtually enhanced perceptions of the interdependence of plant and animal, the collective hoped to inspire viewers to "address the challenges facing our planet in the twenty-first century,"[49] including climate change and deforestation.

As a scholar of speculative fiction, I was intrigued by what I saw as the exhibition's appeal to that subject in its staging. The exhibition's implied narrative, in which ordinary people are invited to use science fictional technology in order to "travel" into fantastic regions, recalls the kinds of imagined voyage tales made newly popular in the nineteenth century by Jules Verne. In its virtual voyage into aerial ocean, *We Live in an Ocean of Air* thus seemed to offer a rare opportunity to think through the legacy of eighteenth- and nineteenth-century representations of air for the twenty-first century.[50]

But what stood out about my visit to *We Live in an Ocean of Air* was its signaling of a change in cultural imaginings of air. Reviews of the 2019 exhibit made it clear that the sight of fellow visitors interacting with their breath around the room was one of the highlights of the experience: "an artwork within an artwork."[51] But by 2021, the pandemic had transformed the VR exhibit into a site where people could model the potential spread of the coronavirus in that very room. Far from blindly swimming through an immersive forest experience, the Montreal participants stepped out of the way of their fellow visitors' breath flows, trying to avoid the kinds of intra-human ecological connection that the VR made visible. Even the programmer's decision to render human bodies as clusters of red "carbon dioxide" dots seemed now to highlight the potential danger posed by the exhaling bodies in the room. The exhibit was neo-Romantic, true, but in a way that recalled the atmospheric unease of Mary Shelley's *The Last Man* (1826) rather than the wondrous sense of ecological connection articulated in Wordsworth's poetry. As visitors' reactions to the 2021 exhibition of *We Live in an Ocean of Air* indicated, the COVID-19 pandemic has fundamentally shifted our perception of art depicting air, in ways that artists have yet to fully process. The exhibition thus left me with questions: What changes will the pandemic bring to the representation of transparent air in speculative media? How will speculative films, games, and novels—media that because of their participation in a science fictional or Weird genre are already bound to certain narrative tropes involving air—reimagine their narratives in the wake of a very conscious shift in atmotechnic perception?

As I write this in August 2022, there are signs that significant changes are being made to certain speculative properties: The digital action-survival game *Forever Skies* (anticipated for late 2022 release) is apparently altering its ecological plotline to account for the cultural impact of the pandemic. The game, which "takes place on a ruined Earth years after a massive climate disaster," asks the player to inhabit the perspective of a "scientist who returns to Earth in an airship" to find a cure for the disease plaguing the space station to which humans have retreated to survive.[52] This game's use of the familiar tropes of Anthropocene disaster, air as resource, a nineteenth-century form of aerial transportation, and disease, have forced the game developers to reconsider its release into a world altered by the pandemic. In interviews, the developers have indicated that while they had always planned to make "falling ill and recovering" a feature of gameplay,[53] they have changed some of the game's narrative elements to reflect their new sense of responsibility regarding the representation of an epidemic.

As with *Forever Skies*, so with a variety of other speculative media. If the pandemic has altered our relationship to atmotechnic citizenship, if we are perhaps more likely now than before to question our entitlement to protection from dangerous air, then we are likely emerging into a new age of aerial representation. As I have discussed in this chapter, speculative fiction has a long history of representing humans' aerial entanglements, capitalizing on its formal impetus to make the invisible visible to raise questions regarding anthropogenic agency and humans' enmeshment in their aerial environments. If, as Sloterdijk suggests, the atmotechnics of the "space station" have become "a key metaphor for the social architecture of the coming world age,"[54] then we might now have a clearer glimpse of the precarities of that age, via speculative fiction and its visions of transparent air.

Notes

1   Peter Sloterdijk, *Terror from the Air*, trans. Amy Patton and Steve Corcoran (Los Angeles and Cambridge, MA: Semiotext(e); distributed by the MIT Press, 2009), 25.

2   Sloterdijk, *Terror from the Air*, 25.

3   Sloterdijk, *Terror from the Air*, 25.

4   Gregers Andersen, "Greening the Sphere: Towards an Eco-Ethics for the Local and Artificial.," *Symploke* 21, no. 1–2 (2013): 138.

5   As various analyses of environmental racism indicate, the perceived "right" to safe air is hardly universal. Rather, it is one of the tenets of what I am calling atmotechnic citizenship, in which privileged bodies are presumed entitled to "safe" air. A key premise of atmotechnic citizenship is the presumption that the harms caused by atmotechnic generation will be displaced onto the bodies of noncitizens—that hydroelectric dams will be built on indigenous land, for example; that the pollution in the United States will drift over the unairconditioned homes of African Americans; and that heatstroke caused by in India will be felt by manual laborers rather than the middle-class inhabitants of airconditioned apartments.

6   Jesse O. Taylor, *The Sky of Our Manufacture: The London Fog and British Fiction from Dickens to Woolf* (Charlottesville: University of Virginia Press, 2016).

7   Amitav Ghosh, *The Great Derangement: Climate Change and the Unthinkable* (Chicago: University of Chicago Press, 2016), 72.

8   Alexis Lothian, *Old Futures: Speculative Fiction and Queer Possibility* (New York: New York University Press, 2018), 15.

9   Heinlein's embrace of "speculative fiction" led to the term being taken up by science fiction writer, editor, and activist Judith Merril, who subsequently popularized the

term with twentieth-century writers and readers. Robert A. Heinlein, "On the Writing of Speculative Fiction," in *The Nonfiction of Robert Heinlein: Volume 1* (Houston: The Virginia Edition, 2001), 221–26, and Elizabeth Cummins, "Judith Merril: Scouting SF," *Extrapolation* 35, no. 1 (1994): 5–14.

10 In the April 1910 edition of *Aeronautics*, for example, Charlemagne Sirch, an electrical engineer, identifies "speculative fiction" as an important site of inspiration for thinking about the issue of "who owns the heavens." Charlemagne Sirch, "A Newly Discovered Right of Way," in *Aeronautics* 6, no. 4 (1910): 123.

11 Oxford English Dictionary, "'speculate, v.'.," n.d., https://www.oed.com/.

12 Mary Wollstonecraft Shelley, *The Last Man*, ed. Anne MacWhir (Peterborough: Broadview Press, 1996), 55.

13 Shelley, *The Last Man*, 151.

14 Shelley, *The Last Man*, 173.

15 Shelley, *The Last Man*, 181.

16 Shelley, *The Last Man*, 181–82.

17 Shelley, *The Last Man*, 182.

18 Siobhan Carroll, "Crusades Against Frost: Frankenstein, Polar Ice, and Climate Change in 1818," *European Romantic Review* 24, no. 2 (2013): 215–16.

19 Shelley, *The Last Man*, 104.

20 Jules Verne, *Five Weeks in a Balloon* (Westport, CT: Associated Booksellers, 1958).; M.P. Shiel, *The Purple Cloud*, ed. John Sutherland (New York: Penguin, 2012).

21 H.G. Wells, *The War in the Air* (New York: Macmillan, 1908), 20–28.

22 Arthur Conan Doyle, "The Horror of the Heights," in *Tales of Terror and Mystery*, Project Gutenberg Etext; No. 537 (Champaign, IL: Project Gutenberg, 1996), 5. https://www.gutenberg.org/files/537/537-h/537-h.htm

23 Peter Sloterdijk and Hans-Jürgen Heinrichs, *Neither Sun nor Death*, trans. Steve Corcoran (Cambridge, MA: Semiotext(e), 2011), 214.

24 Doyle, "The Horror of the Heights," 8.

25 For an overview of the genre, see Roger Luckhurst, "The Weird: A Dis/Orientation," *Textual Practice* 31, no. 6 (2017): 1041–61, https://doi.org/10.1080/0950236X.2017.1358690.

26 Algernon Blackwood, "The Willows," in *The Dark Descent*, ed. David G. Hartwell (New York: Tom Doherty Associates, 1987), 909–43. Algernon Blackwood, "The Wendigo," in *Horror Stories: Classic Tales from Hoffmann to Hodgson*, ed. Darryl Jones (New York: Oxford University Press, 2014), 378–421.

27 Montague Rhodes James, "'Oh, Whistle, and I'll Come to You, My Lad,'" in *Ghost-Stories of an Antiquary* (Freeport, NY: Books for Libraries Press, 1905), 211.

28 James, "'Oh, Whistle,'" 225.

29 James, "'Oh, Whistle,'" 183.

30 For the relationship between objectivity and the abstraction of the observer's body from the environment see Lorraine Daston and Peter Galison, *Objectivity* (New York and Cambridge, MA: Zone Books; Distributed by the MIT Press, 2007).

31 As Elizabeth Miller has recently argued, "the turn to speculative fiction in the era of climate change is not just the critical imposition of another layer of allegory, for there is … a close connection between the upwelling of these genres and the surge in mineral resource extraction, along with its familiar shadow of resource exhaustion,

in the industrial era." Elizabeth Carolyn Miller, *Extraction Ecologies and the Literature of the Long Exhaustion* (Princeton, NJ: Princeton University Press, 2021), 150.

32   See Jan Golinski, *British Weather and the Climate of Enlightenment* (Chicago: The University of Chicago Press, 2007).

33   James, "'Oh, Whistle,'" 185.

34   Andy Weir, *The Martian* (New York: Random House, 2014), *Gravity* (Warner Bros., 2014). *Interstellar, Legendary* ([Burbank, CA]: Warner Bros. Entertainment, 2015).

35   *Avatar: The Last Airbender* (Hollywood, CA: Paramount Home Entertainment, 2006).

36   *The Happening*, Director's cut (Frankfurt am Main: Twentieth Century Fox Home Entertainment, 2008).

37   *Contagion* ([Burbank, CA]: Warner Bros. Entertainment, 2011).

38   Emily St. John Mandel, *Station Eleven* (New York: Alfred A. Knopf, 2015).

39   Sarah Pinsker, *A Song for a New Day* (New York: Berkley, 2019).

40   Mandel, *Station Eleven*, 35.

41   Mandel, *Station Eleven*, 78.

42   Pinsker, *A Song for a New Day*, 86.

43   Pinsker, *A Song for a New Day*, 160.

44   Pinsker, *A Song for a New Day*, 182.

45   Pinsker, *A Song for a New Day*, 198.

46   Pinsker, *A Song for a New Day*, 192.

47   See Denise Petski, "Sarah Pinsker's 'A Song for a New Day' Novel in Works as TV Series From Jason T. Reed Productions," *Deadline*, June 11, 2020, https://deadline.com/2020/06/sarah-pinsker-a-song-for-a-new-day-novel-tv-series-jason-t-reed-productions-1202956848/, accessed July 8, 2022. Information about the COVID-19-inspired delay came via Sarah Pinsker in a personal conversation on October 11, 2021.

48   "We Live in an Ocean of Air | Immerse Yourself in Nature | PHI," https://phi.ca/en/events/we-live-in-an-ocean-of-air/, accessed November 17, 2021.

49   "We Live in an Ocean of Air | Immerse Yourself in Nature | PHI."

50   For more on the Romantic perception of the atmosphere as an aerial ocean, see Siobhan Carroll, *An Empire of Air and Water: Uncolonizable Space in the British Imagination, 1750–1850* (University of Pennsylvania Press, 2015), 119–22. For an account of Britons' changing perception of the influence of the atmosphere on human bodies, see Golinski, *British Weather and the Climate of Enlightenment*.

51   Billie Manning, "Review: We Live In An Ocean of Air," *Culture Calling*, 2019, https://www.culturecalling.com/uk/features/we-live-in-an-ocean-of-air-review, accessed November 17, 2021.

52   Rebekah Valentine, "Polish Studio Far From Home Unveils Next-Gen Sci-Fi Survival Game Forever Skies," January 14, 2022, https://www.ign.com/articles/polish-studio-far-from-home-next-gen-survival-forever-skies, accessed August 1, 2022.

53   Valentine, "Polish Studio."

54   Sloterdijk and Heinrichs, *Neither Sun nor Death*, 214.

5

Respiratory Realism: Elemental Intimacies between "Carbon Black" and *Red Desert*

Jeff Diamanti

> An apprehension of *being in the air* thus means a heightened sense not just of our 'environments,' be they natural, social, urban, cultural, and so on; it actually means going beyond the divide between organism and environment toward a consciousness of our exchanges with it—the ways we breathe it, feel it on our skin, sweat and shiver, notice the smells and changes of the seasons.
>
> Eva Horn, "Air as Medium."[1]

There are virtually no trees in Michelangelo Antonioni's 1964 *Il Deserto Rosso* (hereafter *Red Desert*), which comes somewhat naturally given the film's setting in Raveena—an industrial landscape in the Veneto region of Italy that from water table up to atmosphere is by the early 1960s choked in a cocktail of petrochemicals. Antonioni wanted trees, or more precisely white trees—a *bosco bianco*—which he had because he paid hundreds of workers one night during the filming of *Red Desert* to paint a small forest of pines white. The morning of the shoot, however, the sun was too bright, the greys and whites too chromatic, and so there are no white trees in the film—only a small selection of isolated pines in a sea of fossil fuels. In the written record of the scene, it is nevertheless clear that while the production wasted an enormous amount of capital and labor, and confused more than a few locals, its inconsistency with the quality of light that morning and the Technicolor finish the film would assume—Antonioni's first foray into color, as it were, representing a major departure from his neo- and tragic realisms of the black-and-white 1950s—paradoxically verified the aesthetic project of *Red Desert*. From the perspective of the film, the director's white forest is still *too*

Jeff Diamanti, "Respiratory Realism: Elemental Intimacies between 'Carbon Black' and *Red Desert*" in: *Imagining Air: Cultural Axiology and the Politics of Invisibility.* University of Exeter Press (2023). © Jeff Diamanti. DOI: 10.47788/BRTE9492

colorful, *too* natural, in a landscape where "synthetic materials dominate,"
where industrial chemicals "will end up reducing the trees to nothing more
than antiquated objects, like horses."[2] Resonant with Carolyn Merchant, T.J.
Demos and Jedidiah Purdy's provocation that there's nothing natural about
nature in today's landscape, the idea behind the white forest was to "submit
this old reality, by discoloring it, to a new one, a new reality that is just as
suggestive."[3] The realism to which the film's use of color on the one hand
and aesthetic management of trees on the other refers therefore works by
turning the time of nature—"this old reality"—into an effect of industrialized
landscapes, and the haze of late industry into the medium that holds the
two in suspension. Hence my argument in this chapter, which is that climate
realism—a political aesthetic I helped investigate in the collection of the
same title in 2020[4]—might be thought of as a periodizing term, in addition
to an aesthetic or formal one, and that the petrochemical basis for Antonioni's
"new reality" helps roughly mark its inauguration. The postindustrial atmosphere leveraged in *Red Desert* pervades both the *mise-en-scène* of the film
and indeed its character's toxic affects. But what I will suggest more specifically here is that the politics of an atmosphere rendered metonymically
economic and ecological anticipates both antiracist environmental theory and
the interpretive challenge posed by anthropogenic nature by some years, and
that the worker struggles informing the aesthetic milieu of *Red Desert* should
be understood as central to its capacity to anticipate our own.

Respiratory Realism

The conjunction of "climate" and "realism" is meant to draw out the aesthetic
challenge of holding the elemental force of macro-ecological dynamics to
existing forms of representation, especially when the medium relating these
two conceptual abstractions is the air we breathe. It is also a provocation to
revise the political propositions underwriting earlier periods of realist theory—
to wit, that the collective consciousness latent in the determinisms of a totality
such as industrial capitalism requires formal mediation for its politicization—
now with the interactive (and irreducible) totalities named by late capitalist
modernity, on the one hand, and anthropogenic climate change, on the other.
This interactive locus of atmosphere's economic and environmental determi-
nations has long been recognized as central to racial capitalism and
neocolonialisms, as recent work by Françoise Vergès and Max Liboiron
demonstrates.[5] In the collection *Climate Realism* and special issue of *Resilience*,
however, my co-editors and I invited scholars from a range of humanistic
orientations to spell out the affordances and limitations of the new horizon
of aesthetic and political realism today. My aim here is to build on that
collective scholarship and to think specifically about the particulate matter
of atmospheric intimacies between the late 1960s moment of rapid industrialization in the Italian North (significant because, as others have argued,
this process came so late as to virtually coincide with *post*industrialization in

Europe more generally) and more recent attempts to take the elemental composition of airborne cloud forms as the medium by which our corporeality and our historicity are toxically intimate. But rather than argue that the new atmospheric aesthetics is a radical break from the postindustrial materialism of the 1960s, I want to instead consider the two as bookends to the emergence of climate's doubled concept: a permeation or respiratory dialectic between the historicity and elemental alterity of a world held to capital's chemical bonds.

My emphasis on permeability is meant to underscore the real and intoxicating intimacies carried into and through respiratory rhythms—intimacies that do not exclude constitutive and codependent entanglements, but are also not reducible to them. Permeation draws seemingly remote and discrete bodies and entities into very literal relation, and often in empirical patterns that are familiar only at the level of the body's aesthesis, or sensory knowledge. To breathe in and to breathe out is to take in what another has worked chemically in their veins, often with a signature of those molecular itineraries. While breath might not function in quite the same way as the mnemonic order of an archive, it certainly retains the ongoing metabolization of its bonds. In toxic or virological terms this goes without saying, but it is just as true of banal or colloquial exchanges: Each breath brought into the body is a reciprocal act, borrowing what bodies in the sea and in soil have worked over, and in return elements are fixed into the biomass of so many forms of life. What permeates our mitochondria and redux reactors, the aerobic and anaerobic expenditures, and aromatic attunement to a situation (in a place and the historicity of that place) is atmosphere. Permeability in this way names the blurring of entity and milieu—your and my body along with the diatoms in the sea and the plants blooming in spring are indeed formally and materially distinct, but their interrelation is neither metaphor nor a priori ethically conceived. They are tacitly interactive, embodying the atmosphere as a condition of their participation in it.

And as elemental thinkers as varied as Derek McCormack, Jean-Thomas Tremblay, and Michelle Murphy insist,[6] atmospheres are not abstract or remote in the way that the meteorological gaze enframes them, but are instead attenuated and agitated, never redactable from "an environmental milieu in which forms of life or entities are conditioned and immersed, one whose variations, taking place at myriad scales and degrees of intensity, [and] can sometimes be felt within and across bodies."[7] In my own critical reading of the elemental, this involved and indexical intimacy with milieu is just as importantly historical, which is to say—in keeping with Tobias Menely's reading of "Anthropocene Air"—an intimation of "the delayed and disaggregated effect of fossil capitalism."[8] Hence my attention to the bookending of an aesthetic period of industrial and ecological permeation—very literally the becoming elemental of late realist film in the Italian North and the becoming industrial of atmospheric aesthetics in the respiratory realism of Anaïs Tondeur's "Carbon Black" series (2018). What is suspended in the aeolian currents of airborne densities is in noncontemporaneous relation to its source—a time lag out of joint with the valences of *de jure* and *de facto*—which is why the forensic jurisprudence otherwise solicited to make a case for climate crimes and environmental injustice is often so difficult to advance empirically.

I am trying to think of these two cultural objects as both bookends to a sociocultural sequence—a sequence that my colleagues and I have tried to capture with the term "climate realism"—and also as in respiratory, affective, and conceptual intimacy with one another: not identical for their intimacy, but indeed as dissonant as two drawn together on vectors and reason that need never resolve into identity. Still, the respiratory realism of *Red Desert* and "Carbon Black" is indexical, by which I mean they are both working over the indexicality of airborne historicity to cultural form itself. In Tondeur's series (2018), the question of how to hold the elemental to form is turned literal, since both the photographic negative and the genre of landscape photography specifically are made permeable to its indexical object. Tondeur teamed with atmospheric scientists to capture the particulate matter present in various landscapes during a multiday hike from the north of the UK to the far south. Each landscape captured in the series is tagged to the respiratory mask she used while moving through its terrain. Using the black carbon captured from the atmosphere on location across the British Isles, Tondeur renders the agent of cloud-condensed chiaroscuro—toxic pollutants carried by currents to communities far removed from the source—into the ink used for the final prints. In this way the photographic image materializes the medium's indexical ontology, placed deictically by particulate matter; and in turn the elemental entanglements holding industrial ecology to permeable atmospheres (affect, aeolian, aesthetic, toxic, and otherwise). To stand on location with her camera is also to be placed by the particulates permeating her lungs in tandem with the shared breathing of soils and grasses and exhaust pipes bonding gases to chemo-affective corporealities. The atmosphere is thick: humid and hot in summers, but dense in the cooled suspension of carbon black in the winter months. What stands out across the series is the deep chiaroscuro binding thick clouds, architectural objects, and the mostly thin horizon line at the bottom of the frame to the medium-specific project of writing landscape with the literal stuff in the air. Foreboding for its inversion of the atmosphere into a record of emissions, and those emissions as the agent of a public health crisis that drifts far off from its source, "Black Carbon" takes the common matter drifting on aeolian currents as a pressure on what it is that holds the landscape photograph to its referent. The chiaroscuro is dramatic but it is the opposite of pathetic fallacy. In "Carbon Black," the mood of our ecological present is concentrated into both the object and medium of representation (see Fig. 5.1).

But the slow violence of respiratory permeations between industrial producers and corporeal communities is nevertheless there, suspended in midair: Indeed, *that violence is the air.* Accumulated and circulated through

Fig. 5.1. Anaïs Tondeur, Shibottle, June 1, 2017, Carbon Black Level (PM2.5):
6,75 µg/m³, Carbon ink print.

hundreds of cycles and planetary rotations, rooted into capillaries and shared
moods between botanical, mammalian, and so many other kinds of bodies,
the exhaust of late capitalist modernity asks for a kind of realist orientation
that begins with what Achille Mbembe calls the reclamation of "the lungs
of our world with a view to forging new ground."[9] But in a strong sense,
this call from Mbembe and the critical tradition he is summoning to make
the case for a respiratory real to politics—a call that I take as epistemically
grounding—is contingent on the aesthetic forms available to permeate the
figures brought to bear on this politics: the particulate and particular
exchanges of breath to air, and air to the cloud-based densities of the industrial heritage.

Postindustrial Clouds

On this point, I want to argue that the permeation of noxious gases, fog,
and moody toxins into the *mise-en-scène* of Italian realist cinema anticipated
this blurring of economy and ecology at the moment industrialization manifested as a mode of working over not just labor and spatial orders but also
the historicity of an atmosphere. In 1964, at the tangible edge of industrialization and postindustrialization in the Italian North, atmospheric media
permeates the lungs and language of those sited by its chemo-affects but
with a distinct sense of historical transition.[10] Antonioni's *Red Desert* opens
where *L'Eclisse* from two years earlier left off—that is, at the interface between
the architectural modernism of brick and mortar realism in the earlier film,

and the atmospheric, cloud-based media that will permeate every scene of the latter. Eventually, the film will find its plot in the neurotic unravelling of Giuliana, played by Monica Vitti, the wife of Ugo, a petrochemical magnate, as well as the object of desire for Corrado, played by Richard Harris, and the mother of Valerio. The film's story is prompted by the macrostructural shifts on the horizon, figured on the ground by the workers on strike outside the factory walls and Corrado's purpose in Raveena, which is to recruit engineers and workers to start a new plant in Argentina. Corrado's Argentinian agenda comes across as a strategy to overcome his own depression—to get out of town, as it were—and yet the plot line doubles as a periscope onto the historical conditions of Italian industry, and more importantly what is about to happen to it. Giuliana is both on the brink of a breakdown most immediately because of a near death encounter with a car on the one hand (which we will eventually learn was a suicide attempt) and the much more drawn-out encounter with the psychosocial contradictions of the industrial landscape as well, and it is their mutual state of industrial malaise become trauma that makes the affair between Guiliana and Corrado recursive to other climates at work in the setting. Economically, Italy's famous *il miracolo economico* was about to enter its best decade of value added per labor unit, an era of productivity that somewhat paradoxically laid the seeds for the crash in the rate of profit, unprecedented (and unabsorbable) labor militancy, and the massive relocation of industrial production to the Global South that would define the 1970s. The film's dramatis personae will come into focus gradually, but only after an opening sequence that moves through steam, haze, and chemical clouds with vague outlines of industrial structures moving through the frame. Finally, a few minutes into a series of blurred out pans, we get focus and some kind of provisional clarity when the camera stops on a flare stack beating into a teal sky, an aesthetic mirror to the thick clouds that will in a moment punctuate the film like paragraph breaks (see Fig. 5.2).

The flare stack here marks the point at which the inside and outside of the factory interact, but also the scene for the explosive transformation of elemental media from fossil fuel to atmospheric form, a change in state that is mediated here very literally by industrial production. The factory in its flare stack, though you only ever see an early instance of this process, is pumping out both aesthetically and environmentally a climate—an atmosphere of off green and dull yellows, greys, and rusty reds. But as we will see in a moment, when the film's female protagonist emerges confused and jittery out of the haze with her young son, butting up against the current of workers on the picket line—the very labor force that Corrado is about to replace with cheaper labor in the Global South—this atmosphere is an affective one too, replete with structures of feelings, neuroses, and confused ideas about labor, gender, and desire on the brink of the postindustrial reconfiguration of regions such as Italy's industrial North. And the interactivity of aesthetic, environmental, and affective atmosphere is what I am nominating here as the cloud-based respiratory realism that Antonioni's shift into color, and

Fig. 5.2. *Red Desert* (dir. Michelangelo Antonioni, 1964).

shift away from black and white neorealism of the 1950s, helps to coordinate. The white forest that didn't make the cut is an aestheticized bio-indicator, and the industrial figure of the chemical cloud will help the film's affective and dramatic structure pivot between depressive malaise and ludic investment in the surplus that exceeds commodity and chemical output—a sort of gross, captains of industry sex drive doubling as joy at the coming postindustrial economy.[11]

Thinking *Red Desert* and "Carbon Black" as materially and aesthetically contemporaneous with one another, though separated by nearly half a century, is meant to be provocative. Clearly, they are different media, cultural geographies, and environments. But they are both working with the modes by which an economic mode of production draws multiple modalities and materialities into real relation through abstraction. The particular (here, the particulate matter, molecules, gases) carried from smokestack to respiratory tract is spatially and temporally particular, but winds scramble the bounds of what marks situatedness as so literally cartographically and culturally unique. At the other end of the postindustrial sequence that begins in the 1960s, the media system we call the cloud, that is, the data storage cloud—and the world of information and communication technologies that now rely on it—come to form the infrastructure of global culture and commerce. We might signpost this shift into a climate realism as a convergence of atmospheric and economic ambiguation. In the recent turn to materiality in media, communications, and environmental studies—scholarship in particular by Nicole Starosielski, Mél Hogan, and Allison Carruth—researchers and activists have found new ways to see the connection between electronic clouds and climate change. The connection here between the infrastructure of cloud-based computing and deindustrialized labor is the energy content modulating both. For instance, according to the US Department of Commerce, per worker consumption of electricity increased by 232 percent between 1950 and 1984.[12] Per worker, the postindustrial is far more energy intensive than the industrial, not just because it relies on a global logistics and supply chain that has been overwhelmingly dependent on cheap fossil fuels, but because the very environment of postindustrial production and consumption is lit up with ever increasing amounts of energy-intensive intelligence. Cloud-based computing is enormously resource intensive, from the ground up through assembly, distribution, maintenance, and operation, yet amid the computational sublime of the cloud is the global division of labor, from the hyperexploitation of miners and service workers around extractive industries to the hyperattention required of postindustrial work. We are increasingly familiar with these figures, but they bear repeating: The cloud today consumes 3 percent of global electricity, or roughly 1.5 times more than the entire UK.[13] Every four years its energy requirements double, and with it the lion's share of postindustrial profit: In Q4 2015, Amazon Web Services (AWS) generated US$7.88 billion in revenue, up 69 percent from the same quarter the year before, and its hardware is only one collection in a rapidly expanding sea of data storage providers, such as EMC (now Dell), IBM, Microsoft Azure, and Google.[14] In 2021, AWS recorded a record US$61 billion in revenue.

Today, the ubiquity of media, which is a result of, but also the condition for, the ubiquity of an industrialized energy system, starts to refigure the forms and flows of labor that keep profit rates competitive. One can plot, in other words, a loop of labor through what Benjamin Bratton calls *The Stack*[15]—the layered reality of hard and soft infrastructures that connects earth, cloud, city, address, interface, and user. Figured vertically or horizontally—from subsoil up to atmospheric waves, or from the gentrified core of the new economy across to the special economic zones of the global economy—the layers that make up the stack pull together this doubled sense of atmosphere as both environment and structure. Industrial work generated one set of experiences in relation to the machine and the factory, and these persist of course at certain levels of the stack. Meanwhile, postindustrial work means something fundamentally different. But what precisely does this new environment of cloud-based commerce mean for the experience of work? In Orit Halpern's words, the cybernetic reconfiguration of interface and cognition is by design meant to capture "agglomerations of nervous stimulations; compartmentalized units of an individual's attentive, even nervous, energy and credit."[16] Using Thomas Whalen's 2000 term "cognisphere" to describe this new work environment, Kathryn Hayles emphasizes that "human aware- ness comprises the tip of a huge pyramid of data flows, most of which occur between machines."[17] This parceling out of human input into a flow of information and energy that, at scale, operates independently of any one user or group of users in many ways literalizes the theory of decentered subject foretold by leading postmodernist theorists, such as Gilles Deleuze, Felix Guattari, and Judith Butler. As a feature of cloud-based computation and commerce, however, the experience of decentered subjectivity is entirely bound up with the global infrastructure of fossil fuels—in other words, experience of the cloud is as close to experiencing our global energy system as filling your car with a tank of gas.

Understood this way, the hard infrastructures mattering respiratory realism centralizes the aesthetic practice of mediating what on the one hand Tung-Hui Hu calls the cloud's cybernetic logic of subject and system—its neutrality on the question of who you are, in other words—with on the other hand the cloud's environmental atmospherics.[18] Important for Hui-Hu are the layers of military infrastructure and logistics upon which the cloud's many networks are built, since it is this proprietary and formal genealogy to data processing and storage that signals the shift in sovereignty from discipline to control—that is, surveillance over flows of information that defines the postindustrial structure of power and much of the resistance to it as well. Also true, though, is the layered relation of contemporary cloud-based realities and the energy infrastructures to which both media and military structures are recursive. These layers constitute one another, but don't translate easily across the threshold of user and system. In other words, you can't see rig workers on your phone screen or coal miners in your Google search. At the level of theory, the relation can be posed, if not verified, but the aesthetic challenge of holding these things in focus remains.

We might then consider the cloud as an aesthesis of visual culture—what Timothy Choy and Jerry Zee call "the contents and discontents of modern atmosphere"; that is, the social, environmental, and economic loops that make up a more general structure of the present.[19] In this general structure of the present, the very idea of nature sequestered from human history is so eroded as to occasion a new atmospheric aesthetic. In my reading, Antonioni's intention to "translate the poetry" of factories and their atmospheric emissions in the language of color—to give viewers an experience of "adjusting" to the new industrial reality—is a signal move that anticipates the visual culture of climate change.

Social Reproduction between Climate and Capital

Curious, then, that neither *Red Desert* nor "Carbon Black" are in any obvious way about the bodies typically figured as victimized by deindustrialization in the geographies (here, the UK and the Italian North) or the invisibilized labor of miners where the geology of energy intensive production originates.[20] Instead, both the film and photographic series focalize atmosphere through the permeable interface of distinctly adjacent forms of labor: the reproductive and affective labor of "the boss's wife," on the one hand, and the inherently risk-bearing body of the creative researcher, on the other. Both are exposed to atmospheric harm as an effect of their position in the economy: Giuliana, for her domestic proximity to the factory and its environment, and Tondeur, for the geophysical requirements of field-specific artistic research. Neither is waged in her labor, though both occupy a central position in the larger economic apparatus increasingly skimming from nonwaged forms of creative, sexual, and affective labor. And hence, too, the risks associated with adjacent positionality become more difficult to quantify and hedge against, as would be otherwise in a situation of organized labor where labor power as an industrial force and form of agency can exert itself in the hidden abode of production. The labor politics of creative, affective, and sexual labor asks for more radical orientation toward exposure and solidarity.[21]

Recent critical scholarship on the rise of artistic research in the European and North American context has been attentive to the complementary function of low- and no-waged work in the creative sector and the epistemological exploitation of artists working alongside well-paid science enterprises. But this economic friction has a long history in the worker struggles of the 1960s and 1970s when the gendered character of work in the factory system became a site of organizational and theoretical tension for Marxist feminists. The film's female subject becomes the bearer of this historical recomposition of labor into an atmospheric resource, not by becoming cybernetic per se, but by inscribing the narrative of offshoring—the search for cheap labor in the Global South, as well as the literal offshoring of energy production with the gendered history of reproductive labor that immediately precedes it in Italy. Automated first in the world's industrial centers was not factory labor, but domestic labor—a key cause in the split role of reproductive laborer in the home and productive labor in the factory, increasingly characteristic of the female subject infamously captured by Selma James and Mariarosa Dalla Costa in "The Power of Women and the Subversion of the Community" (1971).[22] Put in terms of the film's plot, Giuliana pulls the gendered division of labor that precedes deindustrialization into the global division of labor that will result from it. The figure of the reproductive laborer in the postwar period is constitutive of the ostensibly later forms of value production that take place at the level of the subindividual across cloud-based commerce—both, as it were, firmly outside the typical sphere or space of production.

Hence, when Giuliana and her bright green coat emerge from the factory's industrial palate (see Fig. 5.3), she too is out of place, literally like a tree, a horse, or "fish who have stomachs full of oil;" and even though the small shop she owns is meant as a personal outlet, she can't figure out what to sell.

Though she's married to the wealthiest character in the film, she is starving—her first lines issue from an exchange of money for a half-eaten sandwich she buys from a picketer. Consuming it in the bushes outside the factory, Guiliana's gaze becomes locked on a simmering heap of burned up coke. Here the figure of the exhausted extends across the divide of human and nonhuman energy, which in key instances will continue to recur in the form of tailings ponds and the coals of a domestic fireplace (see Fig. 5.4).

It is tempting to read this bizarre identification with exhausted energy source as a metaphor for subjective interiority, but aesthetically the film churns out multiple scenes that work against such a distinction. Complementary to this thread of the cloud's exhausted remains, however, is another reading made available in the film that understands the drama of the plot as the aesthetic ground, and the unfolding forms and intensities of clouds as the film's figurative source—a reading that tracks the kernel of what Mark Hansen calls the "atmospheric media" increasingly defining twenty-first-century subjectivity.[23] For Hansen, following Bernard Stiegler, atmospheric media take hold of the subject "beneath the threshold of attention," and thus work more like an environment than an object with which one engages.[24] "Encompassing everything from social media and data-mining to passive sensing and environmental microsensors," Hansen explains,

Fig. 5.3. Screengrab from *Red Desert* (dir. Michelangelo Antonioni, 1964).

Fig. 5.4. Screengrab from *Red Desert* (dir. Michelangelo Antonioni, 1964).

"twenty-first-century media designate media following their shift from a past-directed recording platform to a data-driven anticipation of the future."[25] It's this faculty of anticipation that interests me in Hansen's treatment of atmospheric media, as a feature embedded visually in the filmic prefiguration of this cloud-based reality. In *Red Desert*, the ubiquity and character of clouds, haze, and smog holds both the viewer and the film's characters in a state of suspension, where objects are out of focus or blurred, and the real source of visual clarity is the clouds close to their source—before the process of dispersion. In other words, what I am suggesting is that atmospheric media here is an early instance of those more contemporary media, albeit still in an atmospheric (rather than electric) form, bound as they are to the smokestack, the petrochemical landscape, and the deep inhalation of an economy about to deindustrialize. Take, for instance, Ugo and Corrado's interrupted business meeting, where the question of how to recruit a crew for South America is muted by an explosion of steam that at first holds the two men in a suspended state of curiosity, and then chases them off scene as its heat draws closer (see Fig. 5.5). Over and over in the film, clouds figure as anything but immaterial. Instead, as a medium, their gaseous state is paradoxically what gives the film both its color and mood—its intensive properties—and its dominant orientation by way of volume and extension—an object form. Clouds are both atmospheric and objects in space; environment and structure.

Clouds carry information and intrigue the beholder in distinct ways in *Red Desert* and "Carbon Black," but what I mean to draw into analytic relation here is the interplay between suspension and permeation between the two. Here, Giuliana, Corrado, and Ugo are all held in suspension by

Fig. 5.5. Screengrab from *Red Desert* (dir. Michelangelo Antonioni, 1964).

various forms of gaseous forms—caught off guard by them in scenes where they are otherwise absorbed with forms of work: Giuliana is starving and has finally found a sandwich to eat; Ugo and Corrado are in the middle of a business deal to move the petrochemical factory to Argentina. Tondeur's fieldwork, too, is a physically involved and legislatively sensitive form of work: Her equipment and aesthetic sensibility take her into fields of labor, even if they are calm and calming. The particulate matter held, suspended in clouds drifting across the British Isles, is the exhaust of industrial work, distributed (though empirically traceable) into the elemental folds of landscapes far off. This moment between suspension and permeation is analytically relevant for any critical project in the environmental movement invested in politicizing the experience of climate, since the cloud that catches the viewer off guard is the same medium by which the subindividuated permeation of ecological injustice and violence gets in the body—an issue of colonial violence mapped with precision by Liboiron in *Pollution is Colonialism*.[26] To be suspended before the toxicity of cloud-based media is to be already in a respiratory relation with it.

Conclusion

For me, the question of ontology as a matter of mediation asks for an ongoing philosophical and involved anthropology of elemental entanglements. The atmosphere of late capitalism—involving as it does the permeation of moods, ideologies, revenue flows, infrastructures, and respiratory tracts conjointly— is nevertheless legible in an aesthetic form—not just because clouds, currents, and capital exceed formal capture, but because they are interactive asymmetrically in deictic determination: They are orders of milieu-specificity. Atmospheres carry moods and chemical bonds, and they do so in technical arrangements that are neither identical nor irreducible.[27] Hence, while the bookends I have offered here between 1964 and 2018 are self-evidently unsubstantiated, they are analytically relevant in my argument that the atmosphere composed by the highwater mark of European industry in the late 1960s is very literally in the air today. Aesthetically, I want to reiterate that the permeation of figure and (atmospheric) ground with a density of particles indexes a realism that is both ecological and economic across these cinematic and photographic media that hold this respiratory realism. Neither put much stock in a fated naturalism or a humanist conservationism, even if in both *Red Desert* and "Carbon Black" we can discern an urgent and empirical public health crisis unfolding through air dense with capital's vapid toxicity. But this toxicity betrays any simple notion of situated struggle when "situation" is analytically divorced from historicity. Permeation in this respiratory realism also permeates the increasingly blurry edges of periodization. Anthropocene air, to recall Menely again, remembers decades and indeed centuries of capital's climate. Climate carries the elemental for decades after the point of introduction, mixing in currents both atmospherically and hydrologically. The soil too, as an archive of environmental history, never forgets the slow violence of colonialism and capitalism's material portfolio. All those bodies drawn into and up through milieu-dependent relation are in respiratory intimacy with that history. In Gernot Böhme's formative account of what he called atmosphere and the new aesthetics in 1993, this was already the common reality named by atmosphere, where ecology and economy become entangled, where object and process, subject and system begin to emplot one another spatially. "Conceived in this fashion," he comes to conclude, "atmospheres are neither something objective, that is, qualities possessed by things, and yet they are something thinglike, belonging to the thing in that things articulate their presence through qualities-conceived as ecstasies"—just as they are not produced by subjects, but are translated and fed back by them.[28] Atmosphere thus becomes the common reality of the perceived and the perceiver, yet what makes it the "fundamental concept of the new aesthetics" is the expanded field of cultural production that by the early 1990s was making it "new." At least a decade before relational and weather aesthetics in art, Böhme would tie atmospheric aesthetics to the expanded field of the new economy by claiming it for the field of design (see Fig. 5.6).

Here, then, we can begin to discern the aesthetic contours of the postindustrial obsession with cloud-based computing, which is to say more generally in cloud-based realism, and the common realities held in focus in the climate realism of *Red Desert* and "Carbon Black." The economic injunction during deindustrialization to seek profit, economic energy, and distributed forms of production outside the traditional architectures of production makes the structural atmosphere of the postindustrial recursive to the environmental

Fair Isle Light house	Fair Isle harbour	North Sea	Aberdeen	Stonehaven
23.05.2017	26.05.2017	27.05.2017	28.05.2017	29.05.2017
Level of PM2.5: 2,1 µg/m³	Level of PM2.5: 12,2 µg/m³	Level of PM2.5: 13,8 µg/m³	Level of PM2.5: 2,64 µg/m³	Level of PM2.5: 5,88 µg/m³

Edinburg	Shilbottle	Scotland-England Border	Newcastle-upon-Tyne	Wakefield
30.05.2017	01.06.2017	02.06.2017	03.06.2017	04.06.2017
Level of PM2.5: 8,18 µg/m³	Level of PM2.5: 4,94 µg/m³	Level of PM2.5: 5,16 µg/m³	Level of PM2.5: 6,21 µg/m³	Level of PM2.5: 4,94 µg/m³

Sutton-in-Ashfield	Peterborough	Borough Market	Tate Modern	Folkestone
05.06.2017	06.06.2017	07.06.2017	07.06.2017	10.06.2017
Level of PM2.5: 5,16 µg/m³	Level of (PM2.5): 1,75 µg/m³	Level of PM2.5: 4,56 µg/m³	Level of (PM2.5): 4,85 µg/m³	Level of PM2.5: 1,89 µg/m³

Fig. 5.6. Anaïs Tondeur, "Masks of the expedition" from *Carbon Black*, 2018.

atmosphere of climate in more ways than one. The postindustrial starts in the factory, and even though the factory leaves the city, its atmosphere remains.

Notes

1. Eva Horn, "Air as Medium," *Grey Room* 73 (2018): 6–25. https://doi.org/10.1162/grey_a_00254

2. Michelangelo Antonioni, "The White Forest," in *The Architecture of Vision*, ed. Carlo di Carlo and Giorgio Tinazzi (Chicago: University of Chicago Press, 1996).

3. Antonioni, "The White Forest."

4. Lynn Badia, Marija Cetinic, and Jeff Diamanti (eds), *Climate Realism* (London: Routledge, 2020). https://doi.org/10.4324/9780429428289

5. Francois Verges, "Breathing: A Revolutionary Act," in *Climate: Our Right to Breathe*, ed. Hiuwai Chu, Meagan Down, Nkule Mabaso, Pablo Martínez, and Corina Oprea (Berlin: L'Internationale & K. Verlag, 2022).

6. Derek McCormack, *Atmospheric Things* (Durham, NC: Duke University Press, 2018); Jean-Thomas Tremblay, *Breathing Aesthetics* (Durham, NC: Duke University Press, 2022); Michelle Murphy, "Afterlife and Decolonial Chemical Relations," *Cultural Anthropology* 32, no. 4 (2017): 494–503. https://doi.org/10.14506/ca32.4.02

7. McCormack, *Atmospheric Things*, 196.

8   Tobias Menely, "Anthropocene Air," *Minnesota Review* 83 (2014): 93. https://doi.org/10.1215/00265667-2782279

9   Achille Mbembe, "The Universal Right to Breathe," *Critical Inquiry* 47 (Winter 2021): S61. https://doi.org/10.1086/711437

10   With thanks to Fred Carter for reminding me of The Political Committee of the Porto Marghera Works, "Against Noxiousness" (February 28, 1971) trans. Lorenzo Feltrin, *Viewpoint Magazine* (April 1, 2021): https://viewpointmag.com/2021/04/01/against-noxiousness-1971/, accessed May 1, 2022.

11   Quoted in Peter Brunette, *The Films of Michelangelo Antonioni* (London: Cambridge University Press, 1998).

12   Quoted in Bernard C. Beaudreau, *Making Markets and Making Money* (London: iUniverse, inc. 2004), 100.

13   http://www.independent.co.uk/environment/global-warming-data-centres-to-consume-three-times-as-much-energy-in-next-decade-experts-warn-a6830086.html.

14   http://www.forbes.com/sites/louiscolumbus/2016/03/13/roundup-of-cloud-computing-forecasts-and-market-estimates-2016/#3cc4e5c74b07, accessed January 3, 2017.

15   Benjamin Bratton, *The Stack* (Cambridge, MA: MIT Press, 2015).

16   Orit Halpern, *Beautiful Data* (Durham, NC: Duke University Press, 2015), 249.

17   Kathryn Hayles, "Unfinished Work: From Cyborg to Cognisphere," *Theory, Culture & Society* 23, no. 7–8 (December 2006): 161. https://doi.org/10.1177/0263276406069229

18   Tung-Hui Hu, *A Prehistory of the Cloud* (Cambridge, MA: MIT Press, 2015).

19   Timothy Choy and Jerry Zee, "Condition-Suspension," *Cultural Anthropology* 30, no. 2 (2015): 210–23. https://doi.org/10.14506/ca30.2.04

20   For a critical framework attentive to the historicity of that erasure, see Kathryn Yusoff, *A Billion Black Anthropocenes or None* (Minneapolis: University of Minnesota Press, 2019). https://doi.org/10.5749/9781452962054

21   See Maya Andrea Gonzalez and Cassandra Troyan, "Heart of A Heartless World," *Blind Field* (May 26, 2016).

22   Mariarosa Della Costa and Selma James, *Women and the Subversion of Community* (PM Press, 2019 [1975]).

23   Mark Hansen, *Feed-Forward* (Durham, NC: Duke University Press, 2014), 17.

24   Mark Hansen, "Foucault and Media: A Missed Encounter?," *The South Atlantic Quarterly* 111, no. 3 (Summer 2012): 497. https://doi.org/10.1215/00382876-1596254

25   Hansen, *Feed Forward*, 4.

26   Max Liboiron, *Pollution is Colonialism* (Durham, NC: Duke University Press, 2021).

27   Delia Hannah, "The Philosopher Against the Clouds," in *Cloud Behaviour*, ed. Nanna Debois Buhl (Berlin: Laboratory for Aesthetics and Ecology & Humboldt Books, 2020).

28   Gernot Böhme, "Atmosphere as the Fundamental Concept of a New Aesthetics," *Thesis Eleven* 36 (1993): 112. https://doi.org/10.1177/072551369303600107

Rumpled Bedsheets and Online Mourning: Social Photography and the COVID-19 Pandemic—Haruka Sakaguchi's *Quarantine Diary* and Marvin Heiferman's Instagram account @whywelook

Corey Dzenko

We conduct much of our most intimate lives in our beds. We think of them as private spaces—as set asides—for solace, sex, and dreaming. For many of us, we keep what goes on in the bedroom private. This, however, is all too often a one-way street. Our governments, through legal enforcements and the political pressures of cultural and religious ideologies, have long been intruding into our bedrooms. In the 1990s, the artist Felix Gonzalez-Torres (1957–1996) struck back. In what could be considered a response to the AIDS epidemic that killed his partner, Ross Laycock, he launched an analogue firestorm with a series of public expressions of loss and memorialization.

In an elegiac black-and-white untitled photograph from 1991, two pillows top the artist's own unmade double bed. The indentations in both pillows and the mattress just beneath suggest the recent presence of two people, now only marked absences. Gonzalez-Torres used the reproducibility of photography to reach beyond an exhibition's walls. He displayed a large-scale print of this image in one of the Museum of Modern Art's galleries in New York City (NYC), where viewers found a printed map that led to twenty-four billboards of the same photograph across the city. One possible interpretation

Corey Dzenko, "Rumpled Bedsheets and Online Mourning: Social Photography and the COVID-19 Pandemic—Haruka Sakaguchi's *Quarantine Diary* and Marvin Heiferman's Instagram account @whywelook" in: *Imagining Air: Cultural Axiology and the Politics of Invisibility*. University of Exeter Press (2023). © Corey Dzenko. DOI: 10.47788/CLTU3625 is that the number of billboards coincided with the January 24 date of Laycock's death (see Fig. 6.1).[1] This display of tangible absence transgresses both the artist's personal privacy and some of the impersonal orthodoxies of Minimalism while calling attention to a public health crisis and its political implications. Similar circumstances resonate with the more recent COVID-19 pandemic in the United States and photographic projects that blur private and public.

Many have concluded that the lack of early federal response to AIDS in the United States was due to the association of the disease with homosexuality and drug use. The AIDS epidemic began to spread in the United States in 1980. Yet it took former Republican President Ronald Reagan (term 1981–1989) five years and thousands of deaths to publicly mention AIDS for the first time when a reporter asked him about the lack of federal funding for research. On April 1, 1987, Reagan delivered his first major speech about AIDS to a gathering of doctors, calling AIDS "public health enemy #1." Although Reagan celebrated the amount of federal money used to combat the epidemic, his own administration asked Congress to cut these funds—it was Congress who expanded federal spending to its fullest amount. The president did not address the public health crisis promptly or adequately.[2]

The Center for Disease Control and Prevention confirmed the first known positive case of COVID-19 in the United States from a January 18, 2020 test. Also politicized, early cases of COVID-19 in the United States spread

Fig. 6.1. Felix Gonzalez-Torres, "Untitled," 1991. Billboard. Dimensions vary with installation. © Felix Gonzalez-Torres. Courtesy of The Felix Gonzalez-Torres Foundation. Installation view of the exhibition "Projects 34: Felix Gonzalez-Torres," May 15, 1992 to June 30, 1992. The Museum of Modern Art Archives, New York. Photographer: Mali Olatunji. © The Museum of Modern Art/Licensed by SCALA / Art Resource, NY.

on the west and east coasts; the first case appeared in Washington State on the west coast. The virus then burgeoned in the metropolises of NYC in the east and Los Angeles in the west. Conservative rhetoric often characterizes these coastal sites as those of the "coastal elite," not so-called real America. Former Republican President Donald Trump's (term 2017–2021) public statements and actions—or inactions—during the COVID-19 pandemic echo the lack of public leadership by Reagan early in the AIDS epidemic. Privately, in early February 2020, Trump called the disease "deadly stuff," yet publicly said it was contained, comparing it to the now-endemic flu, contradicted his own health officials, and called Democrats' criticism of his administration's handling of the pandemic their "new hoax."[3] In the United States, the spread of COVID-19 was not—and is not—just a medical incident or one about the social body based solely on the spread of the virus from person to person. The pandemic has again laid bare various social inequities and bigotries that fall along the lines of the politics of identity and place in the United States. It was—and, at the time of this writing in June 2023, still is—a time of public protests toward more inclusive social justice vis-à-vis rising reactionary hate crimes, proposed and passed legislation, and even an insurrection at the US Capital in denial of the results of the 2020 presidential election, the latter of which all attempt to uphold white supremacist, patriarchal, transphobic, and other oppressive ideals.

Unlike the time of the AIDS epidemic, COVID-19 raged in an era of online tools. Given the respiratory nature of COVID-19 and its spread, many turned to virtual environments to avoid more dangerous spaces with shared air circulation. Such is the case when those who could moved their work and schooling online. For some, the act of "doomscrolling," an obsessive scrolling through disheartening news, contributed to an ongoing sense of grief, in part fueled by online social platforms themselves.[4] Others turned to virtual spaces to process their particular ordeals, inviting viewers to respond in real time. This is the case of two photographic projects from NYC and the nearby region, both initiated early during the pandemic and made public virtually: documentary photographer Haruka Sakaguchi's (b. 1990) *Quarantine Diary* (2020) and curator and writer Marvin Heiferman's (b. 1948) ongoing Instagram account @whywelook (2014–present).[5]

With NYC an early epicenter for the spread of COVID-19 in the United States, on March 20, 2020, then New York state governor Andrew Cuomo declared an emergency executive order effective at 8 pm on March 22. He referred to the ten-point plan as "New York State on PAUSE," or the "Policy that Assures Uniform Safety for Everyone." Lacking federal policies to curtail the spread of the respiratory virus, Cuomo enacted this shelter-in-place order, or "lockdown," for his state as he reported a statewide total of 7,102 confirmed cases of COVID-19, 4,408 in NYC alone.[6] Since then, officials have reported that cases likely numbered higher than initial counts. It is in the context of the early phases of the pandemic that Sakaguchi and Heiferman developed their respective photographic projects.

Photography's role as visual documentarian falters when visualizing COVID-19, a virus invisible in the air. And by freezing a mere fraction of a second, a singular image cannot fully depict an unfolding lived experience, such as the "qualities, rhythms, forces, relations, and movements" that anthropologist Kathleen Stewart locates as part of "atmospheric attunements," or the intensities and generative labor that are part of living in or through things.[7] She writes of a "living through that shows up in the generative precarity of ordinary sensibilities of not knowing what compels, not being able to sit still, being exhausted, being left behind or being ahead of the curve, being in love with some form or life that comes along, being ready for something—anything—to happen, or orienting yourself to the sole goal of making sure that nothing *(more) will* happen." To attend to atmospheric attunements means "chronicling how incommensurate elements hang together in a scene that bodies labor to be in or to get through. In the expressivity of something coming into existence, bodies labor to literally fall into step with the pacing, the habits, the lines of attachment, the responsibilities shouldered, the sentience, of a worlding."[8] To live through the earliest days of the pandemic meant maneuvering in the company of various intensities and unknowns—frequently shifting best public health practices, the destabilizing tensions between those who did what they could to protect their community from this new virus and those who denied its severity, trying to survive exposures along with ongoing individual and social ailments, not knowing when or how life would return to "normal," or knowing that it never would.

As Sakaguchi and Heiferman generated their worlds amid the COVID-19 pandemic's unfolding, they both turned to the structure afforded to them by taking or sharing a photograph a day; their daily photographs and captions provided anchorages in their living through. Single images often fail to tell a complete story. Thus, both composed larger series of images that work together as a whole, and they supplemented their photographs with extended captions. At times, the captions offer descriptive text to contextualize an image. At others, the captions become an additional space for them to attend to their experiences.

Sakaguchi began her month-long diary on March 20, 2020, the day Cuomo announced his lockdown. A history of depression compelled her to take at least one photograph daily for one month to cope with sheltering-in-place while living alone in Brooklyn. She posted *Quarantine Diary* to her website and shared some of her images during her takeover of the *New Yorker*'s photo department's Instagram account a few months later.[9] Heiferman's husband, "trail-blazing cultural historian, writer, and social justice advocate" Maurice Berger (1956–2020), died from COVID-19 complications on March 22, 2020.[10] This was shortly after the couple left their NYC apartment to go north to their other home in New York State's Upper Hudson Valley. Since Berger's death, Heiferman continues to use his already-existent Instagram account, @whywelook, to grieve, initially turning to photography when words failed him. Through the sociability of photography, Sakaguchi and Heiferman opened relational possibilities as their images and texts came to viewers in their own atmospheres—distanced from, but still connected with, Sakaguchi and Heiferman.[11] In this chapter, I will consider the ways in which the use of "social photography" during the COVID-19 pandemic became sites for Sakaguchi and Heiferman to process their personal traumas and grief online publicly, and, in doing so, created a collective environment for viewers to do similar from the safety of their own airspaces.[12] Photography's role in such mourning exposes the conjunction of the social life of photography and the atmospheric attunements of living through the COVID-19 pandemic, as such affects always unfurl among the intersectional politics of identity and place.[13]

I Worry that I Won't Be Able to Get Out of Bed

Haruka Sakaguchi identifies as a "Japanese documentary photographer based in New York City" with clients that include the *New York Times*, *National Geographic*, and *Time Magazine*. When three months old, she moved from Osaka, Japan, to the United States with her parents. She summarizes her documentary work as focusing on "cultural identity and intergenerational trauma."[14] It is important to note that she also works as a web designer, which influences how she presents much of her photographic work. Often her projects extend beyond the singular visual image to juxtapose text of various kinds with multiple photographs. Many of Sakaguchi's earlier projects, both photographic and online, provide precedents for her *Quarantine Diary*.

For instance, to address the ongoing generational impacts of the United States' atomic bombings of Hiroshima and Nagasaki, Sakaguchi combined multiple visual and written components for her *1945* (2017).[15] She took portraits of first-, second-, and third-generation *hibakusha*, or the people affected by the bombings. Her website for *1945* pairs these photographs with handwritten messages from the *hibakusha* to future generations. The project also provides a timeline of events involving Japan leading up to the bombing, audio files of the *hibakusha's* firsthand accounts, and ways for viewers to take action toward the ban of nuclear weapons. In an artist talk, Sakaguchi spoke of learning to embrace her "other job" as a web designer as it helps to fund her photography. Her understanding of web design also extends her work in both form and reach to an audience beyond physical publication media.[16]

From her expanded documentary practice, Sakaguchi used photography to help her get through the first month of the COVID-19 lockdown, which was followed by a phased re-opening of the typically bustling NYC. *Quarantine Diary* includes sixty-five slides that most often alternate between photographic image and dated diaristic text—one each every day in chronological order. This continues her practice of combining text and image online as she has done in many other projects.

Sakaguchi opens her diary with a photograph of her lower legs in bed (see Fig. 6.2). Her window delineates between the infected air outside and the safe space she attempted to maintain in her studio apartment. At times, the private space of her bed became difficult to leave. At other times, it sheltered her as a protective environment apart from the world beyond her windows. In her writing for the diary's first day, she shared: "New York City is officially on lockdown. I feel guilty for being able to work from home during this time of crisis. … I also worry that I won't be able to get out of bed." This text appears after the photograph of her legs. The next diary page shows a photographic view of one of the city's usually busy streets, but the businesses are shuttered. No one is out (see Fig. 6.3).[17]

The following day, March 21, she went to buy groceries. Although the store marked the space for social distancing, an unmasked elderly white man invaded her airspace. He hovered so closely to her that she could almost feel his breath.[18] When she asked him to back away, he insulted her and cut ahead of her in line. She remembered:

> An old man called me a chink today.
> Even worse, I didn't say anything back.
>   I don't know why it still affects me to this day. All of those years I'd spent defending myself out on the playground. All of those years I'd waxed poetic about Asian American empowerment and media representation and showing up for each other.

Fig. 6.2. Haruka Sakaguchi, *April 20, 2020*. *Quarantine Diary* (March 20, 2020–April 20, 2020). https://www.harukasakaguchi.com/quarantine-diary. © Haruka Sakaguchi. Courtesy of the photographer.

Fig. 6.3. Haruka Sakaguchi, *March 20, 2020*. *Quarantine Diary* (March 20, 2020–April 20, 2020). https://www.harukasakaguchi.com/quarantine-diary. © Haruka Sakaguchi. Courtesy of the photographer.

> You'd think I'd have something to say. You'd think I'd be able to show up for myself.
>
> Instead, I shifted my gaze down and continued to stand in line as if nothing happened. I walked home with arms full of grocery bags, of canned goods and toilet paper. I sat on the toilet and cried.

She included a photograph of a plastic grocery bag, a crinkled and discardable object with a smiley face printed on it that mockingly refutes her experience (see Fig. 6.4). This racist harassment of the photographer is not a singular event in her month-long diary.

On March 25, Sakaguchi woke up to the news that strangers had attacked her friend, who is Chinese, in the subway the night before, triggering his PTSD. She wrote: "I spent the rest of the day trying to stay busy and productive, so my mind doesn't wander off to dark places. I'm afraid to go outside." The paired image repeats Sakaguchi six times in a single frame on or very close to her bed (see Fig. 6.5). Her bed, in this instance, offers her a protective space, free from the virus but also free from verbal or physical attacks. On March 31, she confessed: "I couldn't get out of bed today. I have no photos." She uploaded only a black rectangle. The following day, April 1, she caught up on editing and took time lapse photographs: In a grid of sixteen still images, the changing light from outside falls onto her empty and disheveled bed. She lamented: "One annoying thing about depression is that it tricks you into thinking that your fears are unprecedented and that you're so. Damn. Unique."

Sakaguchi took a selfie in bed and shared it on April 2, characterizing the act as self-indulgent while also strangely therapeutic. On April 4 she

Fig. 6.4. Haruka Sakaguchi, *March 21, 2020. Quarantine Diary* (March 20, 2020–April 20, 2020). https://www.harukasakaguchi.com/quarantine-diary. © Haruka Sakaguchi. Courtesy of the photographer.

wrote: "Remember all those feel-good activities that I wanted to do before – like cooking, taking an online course, reading for an entire afternoon? Well, I've done exactly none of those things. Instead I just do this:" Dressed in a tank top and underwear, she has curled into a ball on her bed and extends her arms behind her. The left side of her face sinks into her bedsheets as she looks over to the camera with her right eye. On April 12, she again could not leave her bed, even though friends called her phone to check on her. With her camera, she captured an abstracted, angular exposure of her ceiling.

The only color photograph in the diary shows the photographer nude, except for black underwear, in a more formal self-portrait. Seated on her white bed and against her blank white wall, she bends her knees up to her chest, wraps her right arm around to hold her head, and drapes her left across her knees (see Fig. 6.6). For this photograph dated April 8, about

Fig. 6.5. Haruka Sakaguchi, *March 25, 2020. Quarantine Diary* (March 20, 2020–April 20, 2020). https://www.harukasakaguchi.com/quarantine-diary. © Haruka Sakaguchi. Courtesy of the photographer.

two-thirds through NYC's first phase of lockdown, she reconsidered what she had prioritized thus far:

> I'm becoming painfully aware of how much I've relied on work to define me. I've sidelined everything else – friends, family, hobbies, everything. Why? Because New York City is a place where you come to work, work, work until you're too jaded and tired to deal with the MTA [NYC's public transit system] and hopefully have enough money saved up by then to relocate to a quiet fishing town in Maine [a more rural state to the north].
>
> Now that work has dried up, I have no idea who I am. I am just a barely functioning 30-year-old who lives alone in New York City.

Work emerges as a theme other times in her diary, such as when she accepted her first COVID-19 assignment back on March 23 and photographed her camera bag lying open on a patterned vinyl floor. Some people say she is "an idiot for taking the job," but "my photographer friends understand—if not for the money, for the reminder that I can still be of use in this world." Sakaguchi's professional work also led her to encounter anti-Asian bigotry once again as she recorded in her entry of March 30. When on assignment in nearby Pennsylvania, she had trouble photographing her subject while keeping distanced for their mutual safety. She walked around town "yearning for a poetic encounter" after her uninspiring shoot, when she heard someone drive by and

Fig. 6.6. Haruka Sakaguchi, *April 8, 2020*. *Quarantine Diary* (March 20, 2020–April 20, 2020). https://www.harukasakaguchi.com/quarantine-diary. © Haruka Sakaguchi. Courtesy of the photographer.

yell, "Go back to China!" She reported: "A Ford pickup. Typical. I decided that I had had enough of endangering myself with my face. I drove back home." The day's photograph is a montage of eight selfies inside a car. Masked and gloved for protective reasons, she holds her camera in front of her (see Fig. 6.7). According to the Federal Bureau of Investigation (FBI), anti-Asian hate crimes rose 73 percent in 2020, a disproportionate increase as hate crimes in general rose 13 percent.[19] Days prior to the onset of NYC's lockdown, Trump and others referred to COVID-19 as the "Chinese virus," as the former president

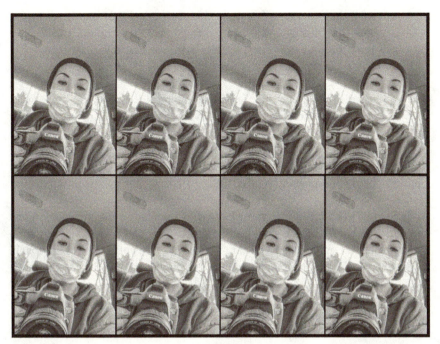

Fig. 6.7. Haruka Sakaguchi, *March 30, 2020. Quarantine Diary* (March 20, 2020–April 20, 2020). https://www.harukasakaguchi.com/quarantine-diary. © Haruka Sakaguchi. Courtesy of the photographer.

continued to do into the years that followed, further encouraging anti-Asian hate crimes.[20] In a 2020 interview, Sakaguchi recalled that she was "already anticipating an increase in hate crimes against Asian Americans" owing to Trump's comments. She wanted to believe her harassment was an isolated incident since she lived in such a racially diverse city. Yet she asked around and found online data collections where Asian Americans self-reported hate crimes they were experiencing with rising numbers that shocked her.[21]

On April 10, she returned to her earlier verbal assaults to further process her experiences. Inspired by an interview in which the writer Jiayang Fan talked of the "probational nature" of her Chinese-Americanness during the pandemic, Sakaguchi shared:

> I have also been wary of calling out racism against Asian Americans because my instincts as an immigrant would snap back, "Well, were you physically hurt? Did they mug/attack/steal from you? Is this how you want to be remembered – a victim?" Then I would juxtapose my problems with the struggles of other communities, and decide that it's trivial, unimportant, unremarkable – I mean, at the end of the day, they're just words, right?
>
> But listening to her recount the terror that she felt when a neighbor repeated racial slurs as she walked by brought tears to my eyes. It

validated the fear that I felt when that man called me what he did
or when that passing truck told me to go back to a country that I
had never been to. The interview reminded me that my experiences
as an Asian American do matter.

The photograph paired with her reflection includes three angled shots of an
above-ground subway track, each progressively closing in to claustrophobically
crowd out more of the open sky. But this is also a scene from out in public
where the photographer connects with others in the larger world. Sakaguchi
has come to realize incidents she encounters in her life are not singular.
Sharing them is important so that individuals do not feel so isolated and
fearful.[22]

Quarantine Diary differs from many of Sakaguchi's other series owing to
the personal nature of the project. She tends to spend a great deal of time
interviewing her subjects and researching for a more collaborative approach.
However, she did use her creative practice to connect her pandemic experience
with those of other Asian Americans. Her series *I Will Not Stand Silent*
appeared in the July 6, 2020, issue of *Time* magazine.[23] For this, she virtually
photographed ten New York-based Asian Americans through FaceTime, a
video and audio chatting platform. Their own handwriting notes their names,
ages, and immigration generations. Extended texts retell firsthand accounts
of what happened; many of Sakaguchi's collaborators recalled being blamed
for the "Chinese virus" and some were misidentified as Chinese, as Sakaguchi
was repeatedly. The photographer returned to the locations where aggressors
harassed her collaborators, reclaiming the sites photographically on their
behalf. She centered the series' portraits on top of the cityscapes she recorded.
The act of photographing thus generatively "turns sites of victimization into
sites of resistance," reviving the agency of those attacked.[24] In some sense,
this project gave the photographer an opportunity to respond to her earlier
harasser at the grocery store; she included a portrait of herself, a photograph
from the store, and a chronicle of what happened, overcoming her prior
silence in the face of her assault.

Sakaguchi and many of her collaborators for *I Will Not Stand Silent* connect
their lived experiences with calls for antiracism at the forefront of the recent
Black Lives Matter movement. For example, two men followed Jay Koo, who
identifies as first-generation Chinese American, while yelling racial slurs and
quoting Trump's talk of the "Chinese virus." They told Koo: "You got the
virus. We have to kill you." Koo said: "I was reminded that the recent attacks
against Asian-American communities due to COVID and the murder of
George Floyd are connected and rooted in racist histories. ... We can never
truly be free unless we are all free."[25] Of her own confrontation at the grocery
store, Sakaguchi detailed:

Before the Black Lives Matter protests, I had contextualized my
incident as of an act of aggression by a single individual—a "bad

apple," so to speak. ... But after witnessing the unfolding of the antiracism movements and encountering heated debates between police abolitionists and those who cling to the "few bad apples" theory, I came to realize that I too had internalized the "bad apple" narrative. I gave my aggressor—an elderly white man—the benefit of the doubt.

She concluded that as an immigrant, she had been conditioned to understand white Americans as individuals. Yet "[t]he protests have brought public attention to the idea that individuality is a luxury afforded to a privileged class, no matter how reckless their behavior or consequential their actions."[26]

Sakaguchi started to share photographs from her *Quarantine Diary* on the *New Yorker* photography department's Instagram feed on May 6, 2020. In this Instagram takeover, she did not reproduce her diary's complete text. She only noted that undertaking her photographic diary helped her get out of bed most mornings. She posted daily until May 12, when she signed off with a diptych of pigeons flying up into the air, thanking everyone for their kind words and for coming along on her "self-reflective journey." She concluded her last entry hoping that all viewers and their loved ones would "stay safe and well." Based on the content of her work, it is clear she meant to stay safe not just from COVID-19, but also from social affronts.

The majority of the public response to *Quarantine Diary* was positive. More so, the project worked. As Sakaguchi explained in the *New Yorker*'s feed, a photo a day turned out to be "the only reason I needed to get out of bed most mornings." In the diary itself, on April 15, five days before the end of the project, she questioned if documenting the past month had accomplished anything. She concluded: "All that matters is, I have yet to spend an entire week in bed."

With a studied understanding of photography's range of expressions, Sakaguchi's diary contains photographs of different genres—abstractions, cityscapes, still lifes, self-portraits, and more informal selfies, showcasing the medium's multiplicity. In *The Social Photo: On Photography and Social Media*, cultural theorist Nathan Jurgenson defines: "Photography is *social photography* to the degree that its central use is more expressive than informational, when the recording of reality is not its own end but a means for communicating an experience."[27] He encourages those who analyze such images to address the social aspects of the photographs, rather than emphasizing only the aesthetics of an image as a strictly formalist methodology of Art History may do. However, Sakaguchi attended to the aesthetics of her images closely while also using the social aspects that stem, in part, from the medium's reproducibility, now in virtual realms.[28] Sakaguchi initially made *Quarantine Diary* for herself while living through NYC's COVID-19 lockdown. As she uploaded her diary to her website and then to social media, her project moved from a personal daily structure to a public presentation for others to openly consider and negotiate their experiences, too. As Sakaguchi has recognized, it was important not just to make but to share her *Quarantine Diary* through photography as a social medium.

No Words Expressed What I Was Feeling

Curator and writer Marvin Heiferman similarly turned to making, sharing, and interacting through "social photos" as part of his grieving process following the unexpected death of his husband, Maurice Berger. Like Sakaguchi, Heiferman pays close attention to both the aesthetics of his single frames and his curation of his Instagram account as a whole. He has organized numerous projects about photography and visual culture for the Museum of Modern Art, Smithsonian Institution, International Center for Photography, and others. Currently he serves as faculty at the School of Visual Art in NYC. In his early career, he worked closely with several artists and photographers who are now largely recognized in the art world and by Art History. He continues to contribute a substantial body of essays and book publications; he is currently developing a book about photography and loss, and has shared his writing process in his Instagram posts.[29]

Heiferman signed up for his Facebook (@Marvin Heiferman) and Twitter (@whywelook) accounts in around 2012–2013 after he had finished a project for the Smithsonian and wanted to put forward his ideas about visual culture. He has always circulated the same material on Facebook and Twitter, and began branding his posts with the question "WHY WE LOOK?" a few years later.[30] Daily, he links to articles about photography and other forms of lens-based visual culture. Even with about 3,900 followers for his Facebook profile and almost 3,200 on Twitter at the time of this writing in June 2023, many of his posts do not garner much, or sometimes any, reaction. Heiferman does not tend to interact with comments in these accounts the way he does in Instagram.

Developers released the more visual Instagram in 2010 to share photos and videos. Heiferman first posted to Instagram on July 29, 2014, when one follower prodded: "What is the world coming to. Marvin Heiferman is on Instagram :)." Another said: "Welcome to Instagram! Looking forward to seeing more of why we look!," referring to the tagline from Heiferman's other social media accounts. His early Instagram posts tend to capture conceptually driven or visually abstracted compositions from what he sees, using the photographic frame to lead his followers to inquire about where or what they are seeing with photographs from public transit, walks, travels, visits to museums, and more mundane acts, such as getting his car repaired or shopping (see Fig. 6.8).

Heiferman joined Instagram because he was part of a research project that required he use the app. At a conference he had seen a few photo directors and publication editors "compulsively taking and posting pictures," and he decided that he should take pictures as well. Decades earlier, he had

Fig. 6.8. Marvin Heiferman, selection of images August 10–15, 2018, posted to his Instagram account, @whywelook (2014–present). In the lower left frame, Heiferman and Berger appear in the red surveillance monitor while using self-checkout at a retail chain store. https://www.instagram.com/whywelook/?hl=en. Courtesy of Marvin Heiferman.

wanted to be an artist, making square paintings and works on paper for some time. He also took photographs while he worked for Light Gallery and Castelli Gallery surrounded by many photographers.[31] Furthermore:

> Because I'd worked with so many incredible photographers—including Harry Callahan, Eve Arnold, Mary Ellen Mark, Robert Adams, Peter Hujar—who shot square pictures, I had a pretty good understanding of what makes square photographic images work, or not. I'd also done a number of curatorial projects for Polaroid, so had access to SX 70s [cameras] and lots of film, so the options and limitations of square-picture-making was something I played around with and was fascinated with.[32]

Such a formal interest in Instagram's square format may seem counter to Jurgenson's emphasis on expression and communication with social photos. But the question of audience and dialog in Instagram piqued Heiferman's interest, too. There in the app's environment he found artists, curators, and writers he knew, and a place where he could converse with a much larger audience than would come through a single exhibition. To Heiferman, Instagram felt "familial," in a way. Then he started to virtually meet people that he did not know beyond the app. He says: "[W]hat I loved about Instagram was that I got to do what I said I always want to do, which is see the world the way other people see it." Yet "people liked the pictures I was making, and I didn't think that much about it."[33]

Early on, Heiferman occasionally shared photographs to celebrate his and Berger's professional accomplishments, such as delivering lectures, publishing books, or attending exhibition openings for shows they curated. Berger, who was white, grew up among the conflicting beliefs of his father, who supported calls for racial justice and the ideas of Martin Luther King, Jr., and his mother, a dark-skinned Sephardic Jew, who showed disdain for Black people. Living in a predominantly Black and Puerto Rican housing project in NYC's Lower East Side, Berger witnessed relentless racism and learned much from his neighbors, who allowed him into their lives.[34] This upbringing, and his study of Art History, led Berger to use his writing and curation to address the construction and maintenance of racialized hierarchies in the art world and broader society.[35] Heiferman came to his related profession from his love of television and movies, and from growing up in Brooklyn, where he frequented the Brooklyn Museum. He accepted a scholarship to Columbia University's film school but left graduate school after a semester. He moved to Manhattan in the 1970s, where he met many painters and artists. When he needed work, he took his first art job at the recently opened Light Gallery—a gallery dedicated to photography—finding employment in the arts in various ways. As Heiferman described, both he and Berger saw the brackets of the fields they addressed in their work and strained against them, shaping their ideas alongside each other's.[36]

The couple met each other not once, but twice. In the mid-1980s, Heiferman held an event with the photographer Barbara Kruger. Because she was nervous, she brought Berger, a friend of hers. Heiferman says he only "vaguely remembered that because he was very tall, skinny, and bearded at that time." Berger developed a crush on Heiferman and engineered that a friend set up a blind date between the two in 1993, which was not a blind date at all for Berger. They had one three-hour lunch and walked out both thinking, "This is it." They married as soon as New York State legalized gay marriage in 2011.[37]

Personal hints of their relationship, home, and interests populate some of Heiferman's earlier posts, such as fragmented glimpses of Berger in the car. Although Heiferman does not geographically tag his posts, he has identified locations in some of his captions—from NYC to upstate towns such as Poughkeepsie, Rhinebeck, and Albany. Once Heiferman changed the focus of his account in the wake of Berger's death, he shared many more photographs of himself, of Berger, and of them together, highlighting them in the past and emphasizing Berger's absence in the present.

Heiferman's audience on Instagram grew considerably since he changed how he uses the platform; in June 2023, he had 14,400 followers, substantially more than his Facebook and Twitter audiences. A May 2022 *New Yorker* article about his display of his grief on Instagram doubled his account's followers in four days.[38] He noted: "The kinds of pictures I was making, though, had changed; they visualized, in one way or another, the very vulnerable state I was in. And to my surprise, they started to resonate as others looked at and commented on them in numbers I was unaccustomed to, with unexpected kindness, empathy and concern."[39] Moreover, "[p]eople tend not to represent themselves on Instagram in this uncomfortable way and what's interesting about this to me is that I'm not only willing but needing to do that."[40]

The COVID-19 pandemic first appeared in Heiferman's Instagram explicitly on March 12, 2020, when he moved online to teach remotely. He wrote: "Talking to grad students over Skype, on the phone, and texting and emailing all day…" with a photograph of his computer screen welcoming him. Around this time, many schools in the US Mid-Atlantic region pivoted to remote course delivery.[41] March 12 is also the day the couple departed NYC for their home upstate, with this date as part of Heiferman's later Instagram narrative, but not one that he posted the day they initially left. For example, on February 10, 2022, two years after the start of the pandemic, Heiferman remembered: "Two surgical masks Maurice bought for us, but left in his night table drawer in NYC when we went upstate on March 12, 2020. I came across them today…." The photograph formally pairs two pink surgical masks lying flat. Their horizontal folds contrast with the black and white striped fabric on which they lie and the more unwieldy curved lines of their white string ear loops (see Fig. 6.9).

On March 13, 2020, the day after the couple left NYC, Heiferman shared "Grim (last night, 10:30PM). Better (today, 2PM)," with two photographs. The first shows a sparsely filled grocery freezer. People asked where this is. Photographer, curator, and writer Michael Lorenzini (@mflorenzini) noted that the grocery store shelves in Harlem—one of NYC's neighborhoods with predominantly African American and Hispanic populations and higher rates of poverty—were stocked full because many there could not afford to stockpile. This commentary draws attention to the ways that economic status impacts the choices individuals have available to them and the geographical division of race and class, even in a single city.[42] The second image reveals a pastoral landscape through a pair of sliding glass doors. When understood with Heiferman's other posts, this view is identifiable as one from his and Berger's Hudson Valley home. Photographer Kari Grimsby (@karigrimsby) celebrated, posting "Glad ur upstate."

The Hudson Valley contains the ancestral lands of indigenous tribes including the Lenape and Mohican, who resided there from time immemorial with some of their descendants still in the region. Following many of their forced removals by the British and US governments, and deaths from diseases introduced by European colonialists, wealthy European American

Fig. 6.9. Marvin Heiferman, "Two surgical masks Maurice bought for us, but left in his night table drawer in NYC when we went upstate on March 12, 2020. I came across them today…" Instagram post (@whywelook), February 10, 2022. https://www.instagram.com/p/CZz9OL9loxm/?hl=en. Courtesy of Marvin Heiferman.

businessmen moved into the Hudson Valley in the early nineteenth century. Prominent New Yorkers then began to collect landscape paintings of the region by Thomas Cole (1801–1848) and other "Hudson River School" (1825–1875) painters.[43] These landscapes circulated as symbols of a Romantic new nationalism, celebrating the natural beauty of the United States, locating the sacredness of God in the landscape, and modifying European traditions of painting. The region now hosts many, many art galleries and events, as art workers like being distanced from NYC, but also appreciate being only a few hours away by car or train.

During the COVID-19 pandemic, the additional development of "Zoom Towns" impacted areas such as the Hudson Valley as those able to leave cities—particularly the wealthier and "nonessential" workers who could work remotely—moved to escape densely populated areas. According to a May 15, 2020 *New York Times* report, in the city's wealthiest neighborhoods,

Fig. 6.10. Marvin Heiferman, "City socks, country air," Instagram post (@whywelook),
June 2, 2019. https://www.instagram.com/p/ByOMQtJgmKG/?hl=en. Courtesy of
Marvin Heiferman.

"residential population decreased by 40 percent or more, while the rest of
the city saw comparably modest changes." Researchers found that those
"driving the exodus do not resemble the city as a whole"; they were "mostly
white in a city that's mostly not" and had "considerably higher" incomes
when compared with residents city-wide.[44] Unlike this recently mobile workforce, Heiferman and Berger purchased their upstate home near friends of
theirs about twenty years ago. [45] They frequently explored the area, showing
their stays in both of their social media accounts. For instance, on June 2,
2019, Heiferman posted a photograph of a pair of black and white, geomet-
rically patterned socks hanging on a wooden fence in front of a lush, rolling
green landscape. He wrote, "City socks, country air," positioning nearby NYC
vis-à-vis the Hudson Valley (see Fig. 6.10). With the start of the COVID-19
pandemic, Heiferman and Berger went north, as they had many times before.

Heiferman posted through March 21, 2020, and then again on March
25, but not on the days in between. He did not announce Berger's death

when it happened on March 22. Instead, the news spread, and many expressed their condolences by commenting on Heiferman's posts from the surrounding days. On March 20, the first photograph in his two-picture album shows four deer eating the winter grass in the couple's upstate yard. An open first aid kit fills the frame of the album's second image. Heiferman captioned: "Nature/Nurture. Deer show up at 8, FedEx at 4." Effie Kapsalis (@digief), Senior Digital Program Officer with the Smithsonian, wrote: "I saw this a few days ago and was relieved you escaped the city. I then heard about Maurice. Send so much love and peace to you. What a loss to everyone, but especially you," with three heart emojis.

The first photograph Heiferman shared when he returned to his Instagram on March 25 shows a US flag wrapped around a tree without a caption. When out for a walk, he saw it and thought, "That's fucked up, and that's how I feel. ... I realized that there were no words that expressed what I was feeling. But I saw this thing that did." Sharing this image began his "photographic shiva," based on the Jewish tradition of grieving communally, but through photographs and socially distanced, online exchanges.[46] Over the next two days, Heiferman added uncaptioned images—a reflection showing Berger's side of their bed and a red flower beginning to bloom. On March 28, he captioned a photograph of closed window blinds "When you wake up alone," the first reference to Berger's death. On March 29, he inscribed a photograph of Berger smiling "Gone..." From here on, Heiferman focused his subsequent posts on remembering the trauma of Berger's death while celebrating him and their love. As he characterized,

> Maurice was very careful in curating his public voice and persona. What's more urgent and important to me is to remember and represent our life together. You don't often see a gay love story like ours represented on Instagram and because of that, too, people took notice and started sending me messages. I'd be sitting in bed late at night and somebody in Milan would be writing to me about the story of his lover's death, how isolated he felt when he couldn't talk about it with people he knew, how he had to grieve alone and how heartened he was to see what I'm doing. A straight friend wrote, "I love my husband but I look at the love you had for Maurice and realized my marriage is a mess." [47]

As Heiferman continued to post, he recalled the past and revealed more details about the end of Berger's life. Posting with a circular reference to time occurs with some frequency in social media, making an occurrence from the past new again. Such is the case with the common hashtag #TBT, or "Throwback Thursday." Yet the delay between Berger's death and Heiferman sharing his photographs later has the additional difficulty of the shock of his devastating loss, more so than a mere desire to make the old new again. On May 9, 2020, he shared a photograph of a helicopter, ambulance, emergency SUV, and the crew who responded to Berger's death on March 22. He wrote: "Once certain images sear themselves into your mind, the challenge is how to live with, neutralize, and/or replace them." In response to some of the comments, he disclosed that he wants to be able to unsee things; it took weeks for him to be able to look at this picture because "the experience of being there and the helplessness I felt were so intense."

On June 22, 2020, Heiferman reported that he had visited Sacco Funeral Home near the couple's house upstate to pick up Berger's ashes. Parallel flat bands in the image show a small, green sliver of grass topped by the funeral home's red brick façade with a cobalt blue door and the building's gray shingled roof. Fluffy clouds float through the airy blue sky in the widest band that tops the image, disrupted only slightly by the top of a tree that extends up behind the funeral home. While there, an employee stunned Heiferman by repeatedly saying that "COVID is a hoax," parroting conservative talking points and conspiracy theories. Many parts of upstate New York, including the Hudson Valley, tend to lean more politically conservative than downstate in NYC. Regions that supported the Republican party of Trump have had lower vaccination rates and higher deaths per capita as conservative propaganda, such as the funeral home employee repeated, denied the severity of the virus.[48] In the post's comments, Heiferman updated that he did receive a very apologetic phone call from the funeral home's director after many people sent messages about the employee's unacceptable behavior during his bereavement. Heiferman was also "stonewalled" by local elected officials when he tried to have COVID-19 recorded as the cause of Berger's death. They refused to swab or autopsy Berger's body even though Berger had symptoms that clearly aligned with the virus.[49]

After some time, Heiferman shared images from the couple's upstate bedroom and the events of Berger's death. On August 25, 2020, Heiferman showed a mundane wooden floor. Yet his caption anchors this floor to a larger meaning: "The place on the floor I did CPR once and have looked at differently ever since," marking this place as one where Heiferman used his breath to try and bring Berger's back. On March 21, 2021, he reflected back on one year prior, explaining more of what happened. Heiferman had been concerned because Berger "always had words for everything" and had woken from "incredible and disturbing dreams" that he could not describe. Heiferman continued with an extended description:

> Throughout the day, I brought him the kinds of food he loved, but he barely ate or drank and never left the bedroom. That evening, I told him I was getting scared and thought we should consider going to the hospital. Maurice didn't want to. He'd been texting with his doctor all day, wasn't having respiratory problems and the local hospital has a poor reputation. He admitted, though, that he was scared, too. And then Maurice said the two of us would get through this together, because that is what we always did.

Fig. 6.11. Marvin Heiferman, "A year ago today and when I went into our bedroom, Maurice was having trouble moving around and his voice was faint. I left, briefly, to go the kitchen and get him a drink. When I got back, something didn't seem right. I screamed Maurice's name, lifted him up, saw the blank look in his eyes. I dialed 911 and, following instructions, did CPR until EMS workers burst into the house, crowded into the bedroom and took over. A Medivac helicopter was called. I followed an ambulance with Maurice in it to a landing site a mile away, where I sat alone in our car, not knowing what was going to happen, until a state trooper approached to tell me Maurice had died. In shock, I somehow managed to drive home. And at some point later on, I stripped the bed, knowing that life—as we had lived, loved and known it—had irrevocably changed." Instagram post (@whywelook), March 22, 2021. https://www.instagram.com/p/CMvFKJzlToi/?hl=en. Courtesy of Marvin Heiferman.

In the corresponding photograph, Berger lies ill in the couple's bed, covered by the bed's comforter with his face and pillow beyond the left edge of the image's frame. Heiferman noted that this is the last picture he ever took of his husband. The next day, March 22, 2021, Heiferman remembered stripping the bed, publicly telling more of the story of Berger's last moments alive (see Fig. 6.11).

Fig. 6.12. Marvin Heiferman, "In the two years since Maurice died, it's only in the last two weeks that I've slept through the night, most nights." Instagram post (@ whywelook), March 28, 2022. https://www.instagram.com/p/CbqVsAgF1RN/?hl=en. Courtesy of Marvin Heiferman.

As he continues his photo-a-day project, Heiferman has shown viewers his bed repeatedly, especially as he had trouble sleeping after his husband died. Finally, on March 28, 2022, Heiferman confessed: "In the two years since Maurice died, it's only in the last two weeks that I've slept through the night, most nights." In the day's photograph, he lies in bed with his eyes closed as morning light streams through the window blinds. His left hand rests on the bed in front of him, the light reflecting from both his and his husband's wedding rings that he wears on his ring finger (see Fig. 6.12).

From the finite number of photographs Heiferman now has of Berger, he has shown historical photographs of both Berger and them as a couple in bed—photographs with changed meanings that now resonate with a combined sense of intimacy and loss. For example, on April 20, 2020, Heiferman posted a picture of his feet near Berger's when watching television in Miami in 2013. Photo curator Rebecca Senf (@beckysenfccp) related: "Having my bare feet near to someone else's signals intimacy. I hadn't realized how much I'd taken it for granted until I got divorced four years ago. I totally get this picture."

Like Sakaguchi, who had taken a photograph a day for the first month of lockdown, Heiferman found solace by posting to Instagram daily. The app, he said,

> gave me a place and structure to sort through what I was experiencing, startling and irrevocable loss, and to shape the heartbreaking sense of absence I was up against into some kind of presence. The picture-a-day regimen helped me confront my own vulnerability and gave me comfort and courage and a platform to share it with others. It helped me clear a path through the weeks and months when I, like others, lost track of time.[50]

Elsewhere, he detailed that "grief is such a disorienting state that I was grateful that instead of spending all day thinking 'How am I gonna get through today?' there'd be a few moments when I'd have to think, 'What picture am I gonna put up today?'"[51] Sakaguchi came to recognize the importance of sharing her experiences virtually, as Heiferman has, too. Through reading about grief, he found that grief needs to be witnessed.[52] This witnessing can aid an individual in their bereavement, but, further, he has drawn strength and learned from "the largely unexplored function that images can serve in the grieving process. What I hope this project continues to do is share images and experiences I know helped me and, I've come to see, can be of interest, use and a comfort to others."[53]

Sakaguchi's project extended beyond the personal mental anguish she contended with during lockdown to highlight ongoing racial divides and bigotries in the United States. Likewise, Heiferman's project also connects to the larger national political climate. As he recounted, grief filled the air during the COVID-19 pandemic. People feared for themselves and their loved ones, not knowing of the coming impacts on their health as well as on their "jobs, careers, homes and their futures. They were fearful of the ideologies and politics that hampered effective response. After the murder of George Floyd in Minneapolis, demands for racial equity and social justice made our vulnerabilities ever more obvious." Heiferman began to make his vulnerability visible, which, in turn, allowed others to similarly share with both Heiferman and members of his online audience.[54]

As the COVID-19 pandemic spiked and spiked again across the United States, it exposed various inequities and discriminations in the nation. Sakaguchi's *Quarantine Diary* and Heiferman's ongoing Instagram account, @whywelook, showcase many of these present fault lines, as well as past precedents, through both what they present in their photographs and the social exchanges motivated by their images. The initial reasons for the creation of their respective series differ: Sakaguchi began her diary to cope with the first days of lockdown and set a definitive end date. However, the public health lockdown is not the only trauma of her lived experience as she continues to address ongoing aggressions and assaults against minoritized people in the United States, particularly those against Asian Americans and Pacific Islanders. Heiferman shifted his already existent photo-a-day practice to find some solace from the sudden death of his husband in the recent past, his online present often reflecting on the past and grappling with ongoing grief. He has arrived at the understanding that his curatorial effort on Instagram is about "photography and love, death, loss, politics, and the pandemic."[55] Both Sakaguchi and Heiferman pay close attention to the aesthetics of their compositions with an acute awareness of photography's various forms. Both also appeal to the aims of "social photography." In turn, both of their photographic projects from the COVID-19 pandemic became of use to a larger community amid the lived experiences of the pandemic's varied atmospheric attunements. Their projects continue to provide a socially distanced environment in which to remember, consider, mourn, grieve, share, and more during a public health crisis and ongoing demands for social justice.

Acknowledgments: In memory of Maurice Berger. Thank you to Haruka Sakaguchi and Marvin Heiferman for their generosity in sharing their experiences through their photography, for the conversations we have had, and for their support of this project. To Tatiana Konrad for the invitation to develop this work and the peer reviewers for their suggestions. To Gay Falk and Dickie Cox for continuing to help me think through my understandings of photography and beyond. To Geoff Fouad for the chance to present an early version of this chapter, and to Chris DeRosa, Kiameesha Evans, Fred McKittrick, Maryanne Rhett, and Rich Veit for comments that helped shape it. This work was supported, in part, by a Creativity and Research Grant from Monmouth University.

Notes

1   MoMA's The Elaine Dannheisser Projects Series showcases artworks by emerging artists, bringing contemporary art to the museum and, in this case, to the city-going public. For *Projects 34*'s brochure, press release, and master checklist of the billboards' locations, see Museum of Modern Art, "Exhibition: *Projects 34: Felix Gonzalez-Torres* (May 16–June 30, 1992)," https://www.moma.org/calendar/exhibitions/368, accessed June 10, 2023.

2   ABC World News Tonight, "President Reagan Delivers First Major Speech on AIDS Epidemic in 1987." Aired April 1, 1987. https://abcnews.go.com/Health/video/president-reagan-delivers-major-speech-aids-epidemic-1987-46492956, accessed June 10, 2023.

3   Lauren Egan, "Trump Calls Coronavirus Democrats' 'New Hoax,'" *NBC News*, February 28, 2020, https://www.nbcnews.com/politics/donald-trump/trump-calls-coronavirus-democrats-new-hoax-n1145721, accessed June 10, 2023.

4   The use of the term "doomscrolling" increased in 2020 with the pandemic's required distancing, but the word can be traced back to at least 2018. See Paige Leskin, "Staying Up Late Reading Scary News? There's a Word for That: 'Doomscrolling,'" *Business Insider*, April 19, 2020, https://www.businessinsider.com/doomscrolling-explainer-coronavirus-twitter-scary-news-late-night-reading-2020-4, accessed June 10, 2023.

5   Haruka Sakaguchi, *Quarantine Diary*, 2020, https://www.harukasakaguchi.com/quarantine-diary, accessed June 10, 2023; and Marvin Heiferman (@whywelook), Instagram account, https://www.instagram.com/whywelook, accessed June 10, 2023.

6   Cuomo's executive order closed all "non-essential" businesses, canceled or postponed all nonessential gatherings of individuals, limited any concentrated gathering of individuals outside their home to workers providing essential services, required individuals to maintain social distancing of at least 6 ft from others in public, mandated that entities that provided essential services help facilitate 6 ft social distancing, limited outdoor recreation to noncontact activities, encouraged individuals to use public transit only when "absolutely necessary," asked sick individuals to only leave home for medical care, asked young people to avoid contact with "vulnerable populations," and encouraged the use of sanitizing materials such as isopropyl alcohol wipes. New York State, "Governor Cuomo Signs the 'New York State on PAUSE' Executive Order," March 20, 2020, https://www.governor.ny.gov/news/governor-cuomo-signs-new-york-state-pause-executive-order, accessed June 10, 2023.

7   Kathleen Stewart, "Atmospheric Attunements," *Environment and Planning D: Society and Space* 29 (2011): 445. https://doi.org/10.1068/d9109

8   Stewart, "Atmospheric Attunements": 452.

9   *New Yorker* Photo Department (@newyorkerphoto), Instagram account, https://www.instagram.com/newyorkerphoto/, accessed June 10, 2023. Sakaguchi reposted this takeover on her personal Instagram (@hsakag), Instagram account, https://www.instagram.com/hsakag/ and showed images from the quarantine on Women Photograph (@womenphotograph), Instagram account, https://www.instagram.com/womenphotograph/, both accessed June 10, 2023.

10   Marvin Heiferman, email message to author, June 3, 2022.

11   Photography's sociability includes how, historically, people corresponded with and gathered around early photographic cards and albums starting in the nineteenth century, used the greater access to snapshot photography to create family albums and slide shows, and now share and comment on images virtually.

12   See Nathan Jurgenson, *The Social Photo: On Photography and Social Media* (London: Verso, 2019).

13   See Clare Hemmings, "Invoking Affect: Cultural Theory and the Ontological Turn," *Cultural Studies* 19, no. 5 (September 2005): 548–67. https://doi.org/10.1080/09502380500365473

14   Haruka Sakaguchi, "About," https://www.harukasakaguchi.com/about, accessed June 10, 2023.

15   Haruka Sakaguchi, *1945*, August 16, 2017, https://www.1945project.com/, accessed June 10, 2023.

16   Sakaguchi's use of expanded documentary form garnered her recognition for *1945* by the Nobel Peace Center. The Center displayed her project for a year starting in

November 2017. Haruka Sakaguchi, "Haruka Sakaguchi: Documentary Photographer," artist talk, Masters in Digital Photography i3 Lecture Series, School of Visual Arts, February 26, 2019, New York City, posted July 30, 2019, 1:02:31, https://youtu.be/jtzZrXMO8Ak, accessed June 10, 2023.

17   The first day of Sakaguchi's diary included a photograph, a diary entry, and a second photograph. The rest of the diary's days present a dated diary entry first, followed by a single image. She photographed the image that starts the diary on April 20, 2020, the last day of her lockdown project. This is how she ended up bookending her diary with similar images and seems to mark day one with two photographs. In general, she took many photographs daily, editing them down to a select image for each day. Haruka Sakaguchi, Zoom interview with author, July 14, 2022.

18   Haruka Sakaguchi, interview by Philippa Thomas, "Coronavirus: Your Stories," BBC News; posted to Haruka Sakaguchi (@hsakag), "BBC World News, 'Coronavirus: Your Stories'," Instagram, July 9, 2020, https://www.instagram.com/p/CCbYgpgDQ6x/, accessed June 10, 2023.

19   Sakshi Venkatraman, "Anti-Asian Hate Crimes Rose 73% Last Year, Updated FBI Data Says," *NBC News*, October 25, 2021, https://www.nbcnews.com/news/asian-america/anti-asian-hate-crimes-rose-73-last-year-updated-fbi-data-says-rcna3741, accessed June 10, 2023.

20   On March 16, 2020, Trump posted on his official Twitter account @realdonaldtrump: "The United States will be powerfully supporting those industries, like Airlines and others, that are particularly affected by the Chinese Virus. We will be stronger than ever before!" Leah Millis, "Trump Tweets About Coronavirus Using Term 'Chinese Virus,'" *NBC News*, March 16, 2020, https://www.nbcnews.com/news/asian-america/trump-tweets-about-coronavirus-using-term-chinese-virus-n1161161, accessed June 10, 2023. For a study of the impact of Trump's tweet, see Yulin Hswen, Xiang Xu, Anna Hing, Jared B. Hawkins, John S. Brownstein, and Gilbert C. Gee, "Association of '#covid19' Versus '#chinesevirus' with Anti-Asian Sentiments on Twitter: March 9–23, 2020," *American Journal of Public Health* 111, no. 5 (May 1, 2021): 956–964, https://doi.org/10.2105/AJPH.2021.306154. This study found that anti-Asian hashtags increased significantly following Trump's March 16 tweet, many implying violence. Furthermore, 50.4 percent of tweets using #chinesevirus included anti-Asian sentiments, when compared with the lesser 19.7 percent of tweets that used #covid19. The experiences of Sakaguchi and many others show that perpetuators often bring aggressions enacted in social media to in-person interactions.

21   Sakaguchi, "Coronavirus: Your Stories."

22   Sakaguchi, "Coronavirus: Your Stories."

23   Sakaguchi published *I Will Not Stand Silent* in Anna Purna Kambhampaty, "Facing Racism," *Time* (July 6–13, 2020), 52–69, which was also made available as Anna Purna Kambhampaty and Haruka Sakaguchi, "'I Will Not Stand Silent.' 10 Asian Americans Reflect on Racism During the Pandemic and the Need for Equality," Lightbox, *Time*, June 25, 2020, https://time.com/5858649/racism-coronavirus/, accessed June 10, 2023.

24   Sakaguchi, "Coronavirus: Your Stories." Extended captions can change the meaning of a photograph of a seemingly mundane place as they get to "the potency of a specific place and to the dislocation of the past's reverberations in the present" once

"the ostensible subject of their photographs is no longer visible, having already passed into history" (108). See Kate Palmer Albers, "After the Fact: Joel Sternfeld's *On This Site*," *Uncertain Histories: Accumulation, Inaccessibility, and Doubt in Contemporary Photography* (Berkeley: University of California Press, 2015), 107–29.

25 Kambhampaty and Sakaguchi, "Facing Racism," 60. Police officer Derek Chauvin, who is white, murdered George Floyd, who was Black, on May 25, 2020, in Minneapolis, Minnesota. Chauvin and other police officers came to a convenience store in Minneapolis after a store clerk alleged that Floyd had purchased cigarettes with a counterfeit $20 bill. During Floyd's arrest, Chauvin kept his knee pinned on Floyd's neck for over eight minutes, even after Floyd lost consciousness, killing him. Widespread public outrage over Chauvin's execution of Floyd led to many public protests, especially because a video captured what happened and because this killing was one of many recent killings of African Americans by police.

26 Kambhampaty and Sakaguchi, "Facing Racism," 61.

27 Jurgenson, 17.

28 In his description, Jurgenson says a "social photo" downplays its *thingness*: "The image object becomes a by-product of communication rather than its focus" (17). In his introduction, he writes that he may make overly sweeping statements about photography to talk about the sociability of the medium. This is one description that overgeneralizes, for it is the specific form of the space of the internet and the collection of images together—their specific, albeit virtual and collective, "thingness"—that makes their communication possible. It is the very nature of the virtual thingness of these presentations of photographs that created a space for people to feel a sense of community from a distance during the COVID-19 pandemic, when public health protocols did not allow for gathering in shared physical space. Still Jurgenson's definition of the "social photo" may further our understanding of Sakaguchi's diary and Heiferman's Instagram. For a related discussion about mythologizing new media projects as "immaterial," see Christiane Paul, "The Myth of Immateriality: Presenting and Preserving New Media," in *Media Art Histories*, ed. Oliver Grau (Cambridge, MA: MIT Press, 2003), 251–74.

29 For instance, see Marvin Heiferman, *Seeing Science: How Photography Reveals the Universe* (New York: Aperture, 2019), and Marvin Heiferman, ed., *Photography Changes Everything* (New York: Aperture, 2012).

30 Marvin Heiferman, email message to author, July 1, 2022.

31 Marvin Heiferman, Zoom interview with author, June 8, 2022.

32 Marvin Heiferman, "Marvin Heiferman on Photography, Love, and the Loss of Maurice Berger," interviewed by Roula Seikaly, Humble Arts Foundation, February 8, 2021, http://hafny.org/blog/2021/2/-marvin-heiferman-on-photography-love-and-the-loss-of-maurice-berger, accessed June 10, 2023.

33 Heiferman, Zoom interview, June 8, 2022.

34 Maurice Berger, "Using Photography to Tell Stories about Race," Lens: Photography, Video, and Visual Journalism, *New York Times*, December 6, 2017, https://lens.blogs.nytimes.com/2017/12/06/using-photography-to-tell-stories-about-race/, accessed June 10, 2023.

35 Berger's groundbreaking early publications include his essay "Are Art Museums Racist?," *Art in America* 78, no. 9 (September 1990): 68–77; his semi-autobiographical *White Lies: Race and the Myths of Whiteness* (New York: Farrar, Straus, and Giroux,

2000); and his exhibition and book *White: Whiteness and Race in Contemporary Art*, Issues in Cultural Theory 7 (Baltimore: Center for Art, Design, and Visual Culture, University of Maryland Baltimore County, 2004).

36 Heiferman, Zoom interview, June 8, 2022.

37 Heiferman, "Marvin Heiferman on Photography."

38 Heiferman, Zoom interview, June 8, 2022. Also see Chris Wiley, "A 'Photographic Shiva' for a Life Partner Lost to Covid," *New Yorker*, May 4, 2022, https://www.newyorker.com/culture/photo-booth/a-photographic-shiva-for-a-life-partner-lost-to-covid, accessed June 10, 2023.

39 Marvin Heiferman, "Documenting My Grief: Living on Instagram in the Aftermath of COVID-19," The Longest Year: 2020+, *Literary Hub*, March 19, 2021, https://lithub.com/documenting-my-grief-living-on-instagram-in-the-aftermath-of-covid-19/, accessed June 10, 2023.

40 Heiferman, "Marvin Heiferman on Photography."

41 This practice was not enacted uniformly around the country as a preventative measure, nor federally mandated. Rather, individual states and even schools each determined how they would modify their practices, if at all.

42 See George Lipsitz, *How Racism Takes Place* (Philadelphia, PA: Temple University Press, 2011).

43 Some histories record the "Hudson River School" as the first "American" art movement. See Richard H. Gassan, *The Birth of American Tourism: New York, the Hudson Valley, and American Culture, 1790–1835* (Amherst: University of Massachusetts Press, 2008); and Greg Young, "The Hudson River School: The Story of an American Art Revolution," *The Bowery Boys Podcast*, May 19, 2022, 1:03:56, https://www.boweryboyshistory.com/2022/05/the-hudson-river-school-the-story-of-an-american-art-revolution.html, accessed June 10, 2023.

44 Kevin Quealy, "The Richest Neighborhoods Emptied Out Most as Coronavirus Hit New York City," *New York Times*, May 15, 2020, https://www.nytimes.com/interactive/2020/05/15/upshot/who-left-new-york-coronavirus.html, accessed June 10, 2023. For the changing population densities upstate, see James Dean, "Pandemic Prompted Exodus from New York City, Gains Upstate," *Cornell Chronicle*, March 24, 2022, https://news.cornell.edu/stories/2022/03/pandemic-prompted-exodus-new-york-city-gains-upstate, accessed June 10, 2023.

45 Heiferman, Zoom interview, June 8, 2022.

46 Wiley, "Photographic Shiva."

47 Heiferman, "Marvin Heiferman on Photography."

48 See David Leonhardt's coverage of "red COVID," including "Red Covid," *New York Times*, September 27, 2021, https://www.nytimes.com/2021/09/27/briefing/covid-red-states-vaccinations.html, accessed June 10, 2023; and "Red Covid, an Update," *New York Times*, February 18, 2022, https://www.nytimes.com/2022/02/18/briefing/red-covid-partisan-deaths-vaccines.html, accessed June 10, 2023.

49 Heiferman, "Documenting My Grief"; Heiferman, "Marvin Heiferman on Photography."

50 Heiferman, "Documenting My Grief."

51 Heiferman, "Marvin Heiferman on Photography."

52 Heiferman cites the work of grief expert David Kessler. See Marvin Heiferman, "Sum of Their Parts," interviewed by Steve Miller, *Musée Magazine*, no. 24, Identity

(2020): 194–205, available at https://museemagazine.com/features/2021/6/10/
from-our-archives-marvin-heiferman, accessed June 10, 2023; and Heiferman,
"Marvin Heiferman on Photography."

53 Heiferman, email message, June 3, 2022.
54 Heiferman, "Sum of Their Parts."
55 Heiferman, "Marvin Heiferman on Photography."

PART III

TRANS-SENSORY AIR:
BODIES AND ENVIRONMENTS

7

Envisioning Experiments
on Air and the Nonhuman

Arthur Rose

The aim of this chapter is to address the historical envisioning of air in some contemporary literary texts and, in so doing, take up Rowan Rose Boyson's challenge to analyze "the *political* meanings of air as both metaphor and as lived reality."[1] I am taking "historical envisioning" to mean the way that the texts cause air to be "seen," as both an object shaped by its scientific history and a vehicle for describing other concerns. I use "envisioning" to highlight the paradox involved in trying to see a substance whose primary quality has always been its invisibility. The resolution of this paradox has often been to look at air's effects: the impact of winds, the creation of vacuums, the addition or subtraction of oxygen in confined spaces. This chapter will be no exception; but it nuances the matter of air's effects to situations that emphasize its use. To see the air, to understand its political implications, we must see it being used.

Historically, experiments that sought to see air in use have often been guilty of using nonhuman life to do so. The birds killed in Robert Boyle's air-pump and the mice in Joseph Priestley's established the roles played by air and, later, oxygen in perpetuating certain forms of breathing life. "[T]he rest of the Creatures were made for Man," Boyle would justify his animal experiments, "since He alone of the Visible World is able to enjoy, use and relish many of the other Creatures."[2] More than 200 years after Boyle wrote these words, Louis Pasteur's experiments into fermentation relied on the actions of vibrios to show that other forms of life flourished in anaerobic environments. Concern for the specificity of that life, however, would have "forced" him "to sacrifice clearness in our work" and "wander from our principal object, which was the determination of the presence or absence of life in general."[3] And when, on the fields of Ypres in 1915, human life became

Arthur Rose, "Envisioning Experiments on Air and the Nonhuman" in: *Imagining Air: Cultural Axiology and the Politics of Invisibility*. University of Exeter Press (2023). © Arthur Rose. DOI: 10.47788/HVZM1065 the testing ground for gas warfare, the effect was to turn the human into something else. A man caught in a chlorine cloud is only "like a man" for Wilfred Owen in "Dulce et Decorum Est."[4] Gas makes "his hanging face, like a devil's sick of sin." Sight combines with sound here, as the listener is urged to "hear, at every jolt, the blood / come gargling from the froth-corrupted lungs," but what haunts the speaker "[i]n all my dreams before my helpless sight" is the image of man plunging at him "guttering, choking, drowning."[5]

Owen's poem raises the possibility that epochal moments in the thinking of air may be captured in literary texts, but it also demonstrates the difficulties of using such texts to establish and reflect upon their own historicity. Gas is a violent producer of obscenity, bitterness, and vile, incurable sores; it is not, nor should it be, a disinterested trope through which to harness Owen's poem as a reflection on air's longer history. When Owen interpellates his reader with appeals to pacing, seeing, and hearing, he implicates that reader; likewise, efforts to envision air historically are implicated in the ethical consequences of that envisioning. To see the air as an object of scientific history inevitably reflects the violence of that history, especially on nonhuman life or human life rendered inhuman. To that end, I turn to recent texts that look back to prior moments in the history of air, and, with the benefit of hindsight, capture something of the violence it entailed. First, I recall Boyle's experiments by considering the Andrew Miller's ekphrastic account of a bird in an air-pump in his novel *Ingenious Pain* (1997).[6] Then I follow Daisy Lafarge's lyrical treatment of Pasteur's experiments on fermentation in her poetry collection *Life Without Air* (2020).[7] Finally, I look to the close conditions faced by Ted Chiang's air-dependent automata in his short story "Exhalation" (2008).[8] In each text, nonhuman life finds itself contained by an experimental structure that dictates its removal from, exposure to, or exhaustion of air. By resurrecting these diverse forms of life, the writers recall the nonhuman suffering that attends the deliberate control of access to air. And yet each text remains partly shaped by its inability to evacuate, fully, the human. To understand the effects of air on the nonhuman remains, even for these texts, a matter for human thinking.

Ingenious Pain

Miller's *Ingenious Pain* tells the story of James Dyer, a man who cannot feel pain or pleasure. Set in the mid-eighteenth century, the novel follows James through his childhood as a huckster's assistant, career as a successful surgeon, and eventual discovery of the ability to feel. The novel's success lies in how it manages this transition: The truncated, opaque approach to James's interior consciousness, focused on observation rather than feeling, that marks the first three-quarters of the book is swept away by a free indirect discourse saturated with a sudden rush of sensation. Long before this moment, however, an adolescent James is kidnapped by Mr. Canning, a gentleman scientist who "is a collector" of human "prodigies," which underscores Miller's "sense of Enlightenment as an 'age of miracles.'"[9] His life at Mr. Canning's house culminates in a demonstration of an air-pump, whose description, mediated by the observations of a still disaffected James, bears a striking resemblance to Joseph Wright of Derby's famous 1768 oil painting, "An Experiment on a Bird in the Air Pump":

> A single strong light stands on the table. Next to it, the complicated focus of the room, is a device, slender at the base, and at its top a shining glass bowl. Inside the bowl is a dove, sometimes still, sometimes beating its wings against the glass. The base of the glass is splashed with the bird's excreta. The gentlemen are gathered around the table. ... Mr Canning stands by the machine holding a handle attached to a pair of leather-cased pistons at the base of the machine. By means of these pistons the air will be removed from the glass. Mr Canning calls the glass bowl 'the receiver' ...
>   "Now, gentlemen," says Canning. He begins to turn the handle. Immediately the bird reacts to the change in its atmosphere. A last wild attempt at flight, to burst the glass. A furious knotted energy. Then an invisible hand settles on its back, pressing it to the bottoms of the receiver. ... Mr Canning turns the handle; the bird is convulsed, its wings half spread, flattened against the glass. The body distorts. Spasms are increasingly marked. Then they weaken to a kind of feeble trembling. The only sound is the steady clicking of the rachets at the top of the pistons. The bird is still. Mr Canning lets go of the handle. There is silence, then the noise of sobbing. ... Mr Canning smiles. He has the face of a wise angel. He ... adjusts a mechanism at the top of the receiver. There is a hiss of air, the bird is instantly revived, though its movements are drunken. Mr Canning reaches in, carefully removes the bird from the glass, cups it tenderly in his hands.[10]

The chiaroscuro effect of the "single, strong light," the convulsing bird in the air-pump, the sobbing girl and gentlemen gathered around a single wise and angelic demonstrator all have their correlatives in Wright's work. The cockatoo has been replaced by a dove, but otherwise the strong resemblance between passage and painting suggests Wright may have had an influence on Miller's description. This is unsurprising: Although Miller does not refer to Wright explicitly, he has described his extensive and extended reading around the eighteenth century to get "a feeling for the period."[11] Using a famous painting from the period provides a compelling source of historical authenticity. In this case, however, Miller's ekphrasis creates an impediment, rather than an aid, to historical accuracy. For what makes this properly ekphrastic, which James Heffernan calls "the verbal representation of graphic representation," is two items of historical inaccuracy that narrow the source of Miller's scene to Wright's otherwise uncited painting.[12]

The "shining glass bowl" and "pair of leather-cased pistons" in Miller's description are an anachronism peculiar to Wright's painting. By the time that a double piston mechanism came to be used in air-pumps of the mid-eighteenth century, Laura Baudot has shown, the glass bowl that characterized Boyle's pump had been replaced by bell jars.[13] Thus, Baudot deduces, "Wright's anachronistic air pump, with its various parts, presents a history of air pump technology, from the first celebrated machina Boyleana to the mid-century double-barreled pumping mechanism used principally for demonstrations."[14] By retaining the glass bowl and double piston mechanism, Miller maintains Wright's anachronism. What we read, therefore, is an ekphrastic account, carefully shorn of its ekphrastic markers and placed in a historical novel as proof of its historical fidelity.

According to Heffernan, "the history of ekphrasis suggests that language releases a narrative impulse which graphic art restricts."[15] The image, in its frozen state, implies a narrative that ekphrastic accounts, like Miller's, explicate. By this logic, the importance of ekphrasis lies in identifying a communion between a static work, caught in a moment freed from time, and a narrative that frees that moment to realize itself and its implications in and through time. This may be why more recent theorists of ekphrasis such as Renate Brosch, Liliane Louvel, and Emma Kafalenos have focused on ekphrasis's performativity and "its interaction with an audience."[16] What, aside from an appeal to historical authenticity, is being performed here? Wright's painting catches the bird at its center in the final moments of its capitulation to the vacuum that surrounds it, but it cannot follow the bird through to the outcome of this event. By taking up Wright's picture, Miller's narrative trades on its tensions, while, by completing its actions, resolving outstanding questions about what happens to the bird, the natural philosopher and his audience. Importantly, the apparent cruelty to the bird in Wright's work is diminished by the obvious care with which it is treated in the Miller text. Or, better, the latent potential for care within the painting is, in Miller's passage, explicated and, by this, returned to one of the descriptions on which it was, in part, based: Boyle's Experiment 41, "Exhibiting several trials touching the respiration of divers sorts of Animals included in the Receiver," described in his 1660 *New Experiments Physico-Mechanical, Touching the Spring of the Air, and its Effects.*[17]

Boyle describes "the experiment of killing birds in a small receiver" where he "commonly found, that within half a minute of an hour, the bird would be surprised by mortal convulsions, and within about a minute more would be stark dead, beyond the recovery of the air."[18] Such experiments, he goes on, were sometimes interrupted by spectators who "rescue" the bird or prevail on the experimenter to let in some air. One such bird was "within about half a minute cast into violent convulsions, and reduced into a sprawling condition, upon the exsuction of the air." But "by the pity of some fair ladies, related to your Lordship, who made me hastily let in some air at the stop-cock, the gasping animal was presently recovered, and in a condition to enjoy the benefit of the ladies compassion." Another was rescued by "a great person, that was spectator to some of these experiments."[19] Glossing this passage, J.J. MacIntosh notes that "the birds' sufferings are not in question, and that they are fit objects for rescue and pity."[20] The ladies that turn away in Wright's painting and the person who sobs in Miller's passage both reflect the pity that, at times, Boyle's spectators would find in themselves. But it also fits into a larger historical point that the air-pump itself represented.

By the mid-eighteenth century, such demonstrations were relatively commonplace: a natural philosopher giving a lecture on pneumatics to interested well-to-do families, by showing how, when an air-pump removes air from a glass container, the vacuum affects living beings. Such experiments were important, Baudot recalls from Steven Shapin and Simon Schaffer, because they offered an alternative to metaphysical speculations about "whether or not nature abhorred a vacuum."[21] "Experimental practices were to rule out of court those problems that bred dispute and divisiveness among philosophers," Shapin and Schaffer write, "and they were to substitute those questions that could generate matters of fact upon which philosophers might agree."[22] Rather than prove, definitively, the nature of the vacuum in the air-pump, such experiments "demonstrate the effects of exhausting air on things and animals and draw conclusions about the nature of air and respiration."[23] By shifting attention away from "why" towards "how," certain factual observations could be agreed upon, even if the metaphysical consequences of these facts remained in dispute.

Crucial to this shift was Boyle's choice of "narrative" to describe these experiments. Narrative helped to focus attention away "from the possible creation of a vacuum" and to the material effects themselves. In Boyle's words,

> I proposed to myself in writing these Narratives, but to awaken the curious to consider and observe what variety of phænomena, in such trials, may be attributed to the season of the year, wherein they are made; and to the strength, bulk, age, peculiar constitutions, &c. that relate to the respective Animal, on which the Experiments are made; besides, what things may on other accounts be fit to be also considered.[24]

Narrative permits Boyle to enumerate the material conditions of the experiment, and the steps taken, without demanding a synthetic claim about its overall significance or that it fit into a larger system. Boyle's experiments on air are simply presented as what air does when observed under specific conditions. Its metaphysics suspended, air, for Boyle, can have material qualities (its "spring") without his having to justify, fully, why it has these qualities. But it achieves this displacement by concentrating on the effects that airlessness has on nonhuman life—the bird—which becomes a visual illustration of its effects.

A parallel may be drawn with Miller's efforts to avoid any definitive explanation as to why James Dyer does not feel pain. Despite a series of correlative circumstances around James's birth and his eventual rebirth into a world of pain—the exceptionally cold winter in which he is born and the actions of the mysterious Mary in creating his awareness of pain—nothing in the narrative ever explains the phenomenon. Indeed, the point of Miller's narrative is to make this explanation superfluous. It is, after all, recasting the eighteenth century as an age of miracles. Narrative, for both Miller and Boyle before him, prioritizes the showing over the telling, "to make," in Joseph Conrad's famous phrase, us "*see*" the effects of air's movements.[25] The air-pump, with its globe of glass, provided a useful device for showing what happened when air was evacuated or, "with a hiss of air," returned to the scene.

Miller uses the air-pump as a symbol for James's own nebulous position as an oddity under scientific observation—as much may be surmised by the presentation of James to the gentlemen soon after its description. James's inability to feel pain (and, by extension, empathy) makes him a chilling focalizer, able to describe both the air-pump and his own subsequent torture in dispassionate, factual terms, with little emotional inflection. Such a character permits Miller to focus on the how of the experiments, at the expense of a why. In so doing, he also refuses to comfort the reader with talk of a greater end to justify these torturous means. Denied a larger purpose, the text forces readers to focus on how entities affected by experiments with air respond, themselves, to the experimental space: a conceit paralleled in Lafarge and Chiang.

Life Without Air

Air's mechanics are easier to observe than are its component parts. The bird may begin to die when air is taken out of the cylinder, but to know why we must have some dim sense of the gas known as oxygen and the key role it plays in respiration. Boyle's narratives ascertain the former by manipulating the devices holding air, while Joseph Priestley, in his "Account of Further Discoveries in Air," finds the latter through different preparations of the air being held.[26] The first ensures the presence, or absence, of a unified object, whereas the second modifies that object's composition. In *The Matter of Air*, Steven Connor observes that, while the mechanical investigation of air had depended upon sequestering air as a single body, "the chemical investigation of air would depend upon another artifice, namely the production of 'factitious airs,' artificial variations in the form of air, that would eventually reveal [its] compound nature."[27] Still, the effects remain a matter of observation. When Priestley produces what he calls a dephlogisticated air (or pure oxygen), he assesses its properties by "introduc[ing] a mouse into it" and observing how "vigorous" it remained after an hour.[28] While these observations advance our scientific and historical understanding of air, it remains formally wedded to an observing third-person narration that aspires to the style of description often found in exemplars of ekphrasis.

To show how this experience can be expressed from within the experimental container, I pivot to Daisy Lafarge's *Life Without Air*. This ambitious collection represents two distinct phases in the development of air's effects: The first parallels pneumatic chemistry's interest in "factitious airs," while the second delves more deeply into the projected experience of organism under observation. This distinction reflects the movement in Lafarge's thinking, for, while the factitious airs do appear in poems from *Life Without Air*, these poems are a residual sequence taken from an earlier pamphlet called *understudies for air* (2017).[29]

understudies for air is a collection of twenty-three lyric poems that use air as a shared conceit to describe particular moods or affective atmospheres. Prefaced by the proposal by Pre-Socratic philosopher Anaximenes that "the source of all things is air," the titles in *understudies* develop variations on air: "desecration air," "asbestos air," "infrastructure air," and so on. The opening poem, "falsification air," presents a nominal thesis statement for the project: "what can I pass on, you ask, / about methods of detecting the air?"[30] The signs, the speaker finds, "are most attracted to states of dereliction" or emptiness that remains "imbued with a residual function."[31] One such example is "the sheets of air / gridlocked in double glazing."[32] What we are given "to understand" is not only the empty spaces where air becomes most apparent, but also that the process of detecting the air is tied to the moderating adjective in the title. Air, in the first poem, is discovered through modes of falsification, which remains a useful method for detecting the airs in other poems but, crucially, not the only method. Each poem detects the air in its own way, as determined by the dominant concern of its title.

Certainly, one can read the titles as moderating a single, underlying substance: air as *arche*, as Anaximenes was purported to think. But this misses the poems' success in deviating from any uniform understanding of air. The emptiness of "falsification air" seems strikingly at odds with the air construed as "parasite" or "a chain-throat / contaminant of life" in "infrastructure air."[33] It might then be better to approach Lafarge's airs as "factitious airs"—artificial variations aimed at diagnosing the different atmospheres that accrue around air in different compound states. This also creates the basis for certain continuities that run from the pamphlet to the larger collection.

By Lafarge's own admission, the earlier poems are "understudies"—"[a] poem that is not quite ready to be a poem but might get called to the stage to act like one"—and might well be read as sketches of a final product. But thematically, poems across the two collections share concerns about how one lives in air. When the addressee of *understudies'* "desecration air" develops "maladaptive / breathing patterns to survive / the air," they anticipate Genie, one of Lafarge's personae, becoming "adept / at the opposite of breathing" in "what Genie got" or the "intense unease and asphyxia" the "she" feels in the collection's coda "under observation."[34] Observation and scrutiny yield other parallels across the collections: between toxic relationships and toxic exposures, which convey the potential for life to emerge in the harshest of conditions. In both, Lafarge turns air into a "lyric substance"—Daniel Tiffany's term for the invocation of "the material substance of things" in lyric poetry.[35] Not unlike the generic "lyric 'air'" that Tiffany alludes to, Lafarge's air is presented as both the source of atmospheres and the medium by which they are carried. This allows Lafarge to describe affective states in material terms, not least the feelings evoked by confining, abusive, or otherwise asphyxiating relationships. These common features suggest the discrepancy between the titles is a mere matter of framing, belying a more basic continuity between the poems themselves. Alternatively, they present the basis for a deeper analysis of Lafarge's development.

To illustrate the latter, I turn now to other poems in *Life Without Air*, whose title comes from Pasteur's *The Physiological Theory of Fermentation*. "La vie sans l'air," Lafarge reminds us, was Pasteur's description of fermentation, the process by which glucose molecules are broken down anaerobically or in an oxygen-free environment. "Through a number of experiments – which included asphyxiating animals to death," Lafarge reminds her readers, "he discovered that some organisms perish from a lack of oxygen, while others are able to thrive in states of airlessness."[36] So life without air is, to be precise, life without oxygen. Ranging from anagrammatic language games to surrealist distortions of reality, Lafarge's poems run the gamut of lyric styles, but they cohere in their Pasteurian tendency to subject her nascent voices and characters to various forms of airlessness. "In this capricious, dreamlike collection," the blurb admits, "characters and scenes traverse states of airlessness, from suffocating relationships and institutions, to toxic environments and ecstatic asphyxiations."[37] But the airlessness these poems describe is not the vacuum's absence, as captured through the historical verisimilitude of Wright's picture or Miller's description. Rather, it is a plenitude, the possibilities that airlessness brings for new forms of anaerobic life. If Jonathan Culler is correct that the lyric "attempt[s] to be itself an event rather than the representation of an event," Lafarge's poems try to secure these spaces of momentary airlessness by being, in themselves, the occasion for experimental thinking.[38]

In the collection's coda, air is construed as a potential interference. The coda is, as Lafarge's notes tell us, based on Part 5 of Pasteur's *Physiological Theory*. Close examination shows it to be a found poem, entirely reconstituted from lines in Pasteur's notes. The opening line of the coda—"In our memoir on Spontaneous Generation"—begins an executive summary of Pasteur's essay, where the corollary is "In our memoir of 1862, on so called Spontaneous Generation, would it not have been an entire mistake to have attempted to assign specific names to the microscopic organisms which we met with in the course of our observations?"[39] Again and again, experimental specifics are deleted: Pasteur's line "On March 23rd, 1875, we filled a 6 litre (about 11 pints) flask, of the shape represented in Fig. 11, and placed it over a heater" becomes Lafarge's "On March 23rd we filled the shape represented."[40] The poem flattens Pasteur's description of depriving the vibrios of air into "[a] most simple method of observing the deadly effect of air," eliminating the objects on whom these methods are acting. While Pasteur writes that "[f]ermentation had become less vigorous without having actually ceased, no doubt because some portions of the liquid had not been brought into contact with the atmospheric oxygen," Lafarge's line—"She was less vigorous without having actually ceased"—reorients the poem away from the lyric subject—the experimenters' collective "we"—towards the action of the vibrios themselves.[41]

Lafarge's strategic deletions focus attention on the fermentation agent, and foreground its gradual becoming active. To exaggerate this effect, Lafarge does not follow Pasteur's own sequence when positioning the lines. Rather, lines are chosen from across the essay to emphasize a clear process of becoming. While the coda broadly follows the experiment that Pasteur details, on the fermentation of lactate of lime in a mineral medium, it also radically reworks Pasteur's formulation into a chronological breakdown of the experiment's stages. As a result, we are prompted to observe the emergence of a new organism, the "she" the poem narrates into being, who lives best "without air," who possesses unusual powers of resistance and so on.

Lafarge's reworking of Pasteur reproduces in practice a theoretical argument advanced by Bruno Latour. In his reading of another Pasteur paper on fermentation, Latour observes two dramas at work in the text, in which the text performs two extraordinary processes of transformation.[42] The first drama is what he calls a Cinderella story, where the fermentation agent, previously a nonentity, is converted into "a glorious and heroic character" through the subtle interventions of Pasteur's opinion.[43] This conversion process is facilitated by a series of ontological transformations: First, the ferment's existence is denied; later, it appears simply as a set of random phenomena, then it begins to assert its own actions, have its own qualities, condition its environment. Pasteur at this point intervenes to liken it to other, similar organisms, only to distinguish it from those organisms and finally place it in a taxonomy. Latour's second drama turns on a reflexive question about who the real agents are: "Who is constructing the facts, who is directing the story, who is pulling the strings—the scientist's prejudice, the nonhumans?"[44] For Latour, Pasteur resolves this question by distributing "activity in the text between himself, the experimenter, and the would-be tentative ferment."[45] In like fashion, we might say that Lafarge's liquid, previously left to herself, gradually assumes more and more of an agentive role, until the experimenters find themselves "forced to regard her as a distinct species."[46]

But to what end? In Lafarge's poetry and her novel *Paul*, there is a preoccupation with the effects of suffocating or abusive relationships, colloquially termed "toxic," on the passive, usually female, partner.[47] Here, this thematic concern finds its analogue in Pasteur's experiments on anaerobic life. Importantly, Lafarge does not simply dismiss such relationships; often they become the basis for thinking about how unexpected forms of flourishing emerge, even in the direst of circumstances. While she does not condone such relationships, she does consider how flourishing happens in spite of and, sometimes, because of them. To isolate this form of flourishing, Lafarge cannot simply invoke a blanket notion of smothering. It demands a notion of air made up of multiple elements: the carbon dioxide that permits an organism to thrive, unimpeded by the oxygen that would kill it.

Lafarge's development can be read as correlative to the move from pneumatic chemistry to its consequences for nonhuman life. If so, however, the paradigms do not simply act as metaphoric sources, providing a pleasant cover for essentially the same theme. By shifting her focus from atmospheres to the laboratory setting, Lafarge translates generic observations about breathing air to the examination of life in anaerobic environments. The abstractions of air as a general constant, given or withdrawn, mutate into the particular forms of air needed to sustain actual lives. What we find in the movement from *understudies* to *Life* is the jettisoning of a conceit that is perhaps too dominant and too confining. Compound air is still used in *Life*, but without controlling the direction of the poems or dictating their conditions.

"Exhalation"

Lafarge's earlier interest in air as the source of all things serves as a useful point of entry into Ted Chiang's 2008 speculative fiction "Exhalation," which opens with the Anaximenesian declaration, "It has long been said that air (which others call argon) is the source of life."[48] That air is "the source" is immediately denied ("this is not in fact the case") in the story's second sentence.[49] The source of life, we will discover, is a difference in air pressure: It is air's movements, rather than air itself, that powers Chiang's narrator.

"Exhalation" is presented as the extinction account of an unnamed metal automaton, engraved on sheets of copper and left to be found by some future civilization. The account details the scribe's efforts to discover the cause of an anomaly: the gradual speeding up of the clocks. This, it realizes, is not because the clocks are speeding up, but because the automata themselves are slowing down. Much of the story goes into explaining why this is the case, as the automaton describes a series of experiments it performs on itself.

The automata live in an atmosphere composed of argon and rely on tanks of the gas in its pressurized form to power their bodies and processors. The processors, as the automaton scribe's experiment uncovers, are made up of hundreds of gold leaves whose gas-powered movements constitute their thoughts. "[A]ir does not," the automaton scribe learns, "simply provide power to the engine that realizes our thoughts. Air is in fact the medium of our thoughts. All that we are is a pattern of air flow. My memories were inscribed, not as grooves on foil or even the position of switches, but as persistent currents of argon."[50] Without the gas, the gold leaves fall into a default position, which doesn't just stop the automata but erases all their thoughts and memories. Hence, they must ensure a constant supply of pressurized gas in their tanks, which they fill at stations linked to a vast pressurized reservoir.

But the automata also live in a closed world. As the tanks release argon, this raises the relative pressure of the environment, bringing it closer to that of the reservoir. As the automaton scribe realizes, the smaller the difference in pressure becomes, the slower the pressurized air moves, affecting not just the automata's movements but their very capacity to think. The pneumatic conditions of life in their world are progressing towards a final equilibrium, where that life can no longer be sustained. The scribe summarizes this entropic movement to a final point of stasis as follows:

> [A]ir is not the source of life. Air can neither be created nor destroyed; the total amount of air in the universe remains constant, and if air were all that we needed to live, we would never die. But in truth the source of life is a difference in air pressure, the flow of air from spaces where it is thick to those where it is thin. The activity of our brains, the motion of our bodies, the action of every machine we have ever built is driven by the movement of air, the force exerted as differing pressures seek to balance each other out. When the pressure everywhere in the universe is the same, all air will be motion-less, and useless; one day we will be surrounded by motionless air and unable to derive any benefit from it.[51]

The story is Chiang's attempt to translate into fiction Roger Penrose's thinking about metabolism and entropy. When we consume food, Penrose argues, we absorb a low-entropy, ordered form of energy and convert it into a high-entropy, disordered form of energy. "In effect," Chiang summarizes, "we are consuming order and generating disorder; we live by increasing the disorder of the universe."[52] Likewise, the ordered, pressurized air becomes, as it is released, disordered, its high entropy an inevitable consequence of its use.

If anything, Chiang's automata suffer from a surfeit of air. Why, then, bring them into this discussion of airless spaces? Because they help to shift my discussion from vacuums and elements to use: Ultimately, the argon-filled space in Chiang's story correlates to both Miller's vacuum or Lafarge's air because the life under observation in all three is "unable to derive any benefit" from the observation, whatever the benefits that accrue to its observers.[53] And yet where Miller and Lafarge's environments are contained and, by extension, act as states of exception, the conditions faced by Chiang's automata are universal to the world they inhabit. Here, we can find a correlative with Peter Sloterdijk's atmoterrorism, which describes "the displacement of destructive action from the [enemy's body] onto his environment."[54] Chiang's detailed description of the environment makes it clear that the automata "become at once victim[s] and unwilling accomplice[s] in [their] own annihilation."[55] This is further clarified by Chiang's separation of the usually entangled phases of inhalation and exhalation: Here, instead of a regular inhalation, the automata swap out tanks of compressed argon. Thus, Jean-Thomas Tremblay observes, "the disembodied inhale literalizes resource extraction, and the embodied exhale a process of extinction coextensive with the achievement of a certain equilibrium or homeostasis."[56] The force threatening the living conditions of the automata is their own cumulative exhalation. The enemy is themselves.

The story presents a thought experiment on the workings of a closed system. Indeed, although the automaton's universe is a cylinder, its workings as a total system have strong parallels with Sloterdijk's larger project on the phenomenology of spheres: his study of the bubble, the globe, and foam as dominant forms of containment. Admittedly, Sloterdijk's *Sphären* are conceptual, whereas, within the frame of the story, Chiang's cylinder is 'real'; nevertheless, both provide morpho-immunological constructs, creating interiors that protect life's possibility from an "outside," while also defining its finitude. Tremblay notes the additional irony of a thought experiment wherein thinking, too, advances species extinction.[57] This may be why the story so successfully translates as an allegory for the role humans play in creating the conditions of their own extinction under the aegis of anthropogenic climate change: The conditions are wrought within a closed system, and thinking has helped make it so.

Some of this is anticipated in the parenthetical observation of the opening sentence: That what is called air in this story is otherwise known as argon. In this story, air is not only not the source of life, but it is also not air as we conventionally understand it. As Tremblay puts it, "Chiang has transported us elsewhere."[58] Gary K. Wolfe has noted that the use of parenthesis recalls Jorge Luis Borges's opening sentence in "The Library of Babel," which begins, "The universe (which others call the Library) is composed of an indefinite, perhaps infinite number of hexagonal galleries."[59] Both parentheses use objects with definite, contextual identity (argon or a library) to qualify terms that otherwise take on universal significance (the air or the universe itself). As a consequence, Wolfe remarks, they "imply a perceptual world outside the circumscribed frame of the narrator's world view."[60] What is universal for the narrator is understood as a definite object for its imagined others. This is why Tremblay is right to find in the prospective "post-thought homeostasis" imagined near the end of the story the possibility of an imagined space constituted by "nothing knowable to Man."[61] Air, argon, "the divine breath ... does not die out ... It exceeds and survives Man; it was never meant to be his alone."[62] The possibility of such knowing, after the end, demands an act of speculation: the creation of a hypothetical other who will observe its effects. Such others, Chiang admits, would have different referents for apparently universal terms (air, universe).

More immediately, the parenthesis marks the later destruction of the automaton's relation to their air milieu as an "unquestionably given and anxiety-free, unproblematic being."[63] When the realities of the air in the automata's world become widely known, it causes "widespread panic," "accu-sations of wasted air," "brawls," and "deaths."[64] Their environment can no longer be taken for granted. In representing the entire world as an experi-mental space, the story shows how returning the air to its environment simply emphasizes that its observers, too, are trapped within its confines. To convey this entrapment, Chiang has his narrator perform a speculative leap beyond "the circumscribed frame of [his] world view": an imagined reception of his text by others, implied in the opening reference to argon and compounded by the possible explorers from other worlds.[65]

Conclusion

Although most read Chiang's story forwards, to a point of anthropogenic extinction, the conditions faced by the automata resemble nothing so much as the bird trapped in Boyle's air-pump or the vibrios in Pasteur's swan-necked flask. We have, by now, switched positions; no longer watching the bird's vain, spasmodic attempts to exit the glass, readers find themselves confined to the vessel at which they once marveled. Against the sense of historical development that inevitably accompanies descriptions of successive experimental phases—here given as the physico-mechanical, chemical, and environmental investigations of air—Miller, Lafarge, and Chiang present alternative points of view from which to observe the effects of such exper-iments: from without, from within, and from a speculative position that runs from within to without and back again.

Each of my examples applies a form of airlessness or, to be more precise, air exhaustion to nonhuman life, the better to imagine human dwelling in constricting situations. As a device, this conceit offers us possible ways for understanding how airlessness may be figured and, by extension, give some sense of what airlessness means to us in the fraught conditions of our contemporary moment: the political meaning Boyson refers to in my opening sentence. At every stage, this is accompanied by a nonhuman being who experiences the airless state on our behalf and, therefore, provokes us to reconsider the fundamental assumption that, when we raise the matter of air, we automatically know what it is we mean. But it also discloses a lingering anthropocentrism even in these works that seek to trouble the limits of the human. As Tremblay notes,

> the ending of "Exhalation" reminds us that Man does not easily relinquish his position. The narrator of the epistolary story eventually addresses the reader as "explorer." ... The generosity of the narrator's address and the miracle of intergalactic recognition barely mask the horror of being hailed as Man. A colonial rationality is projected onto us. We are made to stand as living proof that Man is not done exhaling.[66]

Despite the patient excavation of the vibrios' experience and the ethical refusal to take up their subject position, Lafarge's poem still slips into referring to the ferments as a "she." And, while James Dyer's immunity to pain is, like the coincidence of the bird experiment with his own role as an object of pain experiments, meant to show how he stands at the very limit of what it means to be human, both he and the bird are ultimately recuperated for the human by virtue of the sympathy they elicit. Close as the texts come to attempting to generate a nonhuman perspective, they remain all too human in their appeals to sympathy, their catachrestic conflation of literal and metaphoric suffocation, and their reliance on an ultimate human observer.

Nevertheless, they also provide a counter to a problem raised by Steven Connor, which I turn to by way of closing. Like Peter Sloterdijk before him, Connor's concern is that, in "taking to the air" we have lost our principal "material substrate" for figuring "the immaterial action of thought."[67] Air, Connor concludes his book, "no longer figures the illimitability to which we aspire, but rather the indetermination that we are."[68] Of course, the nonhuman discontents of air's experimental history implicitly call into question Connor's universalizing "we", since, for them, air figures not as a matter of illimitability or indetermination, but as a means of torment. And yet to focus on the existential crisis in his summation misses the attention Connor pays throughout *The Matter of Air* to cataloging air's mediating role. In an earlier presentation, given as a response to contemporary artistic uses of air at Art Basel in 2007, Connor observes: "Air is no longer an ideal image for art, but an object for it to work on, and by which to be itself worked out, worked loose even from its self-identity. ... In propagating the air into objects, art stands a chance of propagating into something beside itself."[69] Whereas once air embodied a world beyond objects, it has become itself an object for art to "work on," a substance that "enters into composition, is folded or forced into new kinds of object."[70]

In the very different texts touched on in this chapter, air is worked on as an object of intellectual history to create the distance needed to understand the text's point of view. This is more explicit in my discussions of Lafarge and Chiang; a marker, perhaps, of the increased sensitivity to the determining role air plays in the ongoing climate emergency. Nevertheless, routing this history through readings of literary works necessarily suggests they do more than simply illustrate popular science. Certainly, the texts gesture towards a common poetics of entrapment, whereby the analysis of air requires a being to be enclosed within it. With this gesture, they force us to ask difficult questions about the points of view we take to the air and the various distances we sustain to it, with concomitant implications for who can breathe and where. But they also demonstrate how points of view are, inevitably, differentiated by focus and genre: Miller's historical novel (following Boyle) shifts our attention from why air works to how, Lafarge's lyrics (following Pasteur) move from atmospheres to the agency of voices within them, and Chiang's speculative fiction (following Penrose) disengages with the substance the better to address its movements. To articulate these processes of change, authors seize both the moment of a painting and its consequences, describe an atmosphere and the voice caught within it, imagine the totality of a world and what lies beyond it. In identifying moments of tension between what is and what might be, these texts allegorize their own efforts to warp air's meaning: allegories of air contingent on our willingness to believe the invisible may yet be envisioned.

Notes

1   Rowan Rose Boyson, "Air and Atmosphere Studies: Enlightenment, Phenomenology and Ecocriticism," *Literature Compass* 19, no. 1–2 (2022): e12654. https://doi.org/10.1111/lic3.12654

2   Robert Boyle, *The Works of Robert Boyle: The Usefulness of Natural Philosophy and Sequels to Spring of the Air, 1662–3*, vol. 3, ed. M. Hunter and E.B. Davis (London: Pickering & Chatto, 2018), 217.

3   Louis Pasteur, *Studies on Fermentation*, trans. Frank Faulkner and D. Constable Robb (London: Macmillan & Co., 1879), 29.

4   Wilfred Owen, *The Collected Poems of Wilfred Owen*, ed. C. Day Lewis (London: Chatto & Windus, 1963), 55.

5   Owen, "Dulce," 55.

6   Andrew Miller, *Ingenious Pain* (London: Sceptre, 1997).

7   Daisy Lafarge, *Life Without Air* (London: Granta, 2020).

8   Ted Chiang, "Exhalation," *Exhalation* (London: Picador, 2019), 37–57.

9   Miller, *Pain*, 119; 133. See G.S. Rousseau, "*Ingenious Pain*: Fiction, History, Biography, and the Miraculous Eighteenth Century," *Eighteenth-Century Life* 25, no. 2 (2001): 58. https://doi.org/10.1215/00982601-25-2-47

10 Miller, *Pain*, 128–30.

11 Andrew Miller, *Ingenious Pain* (Lancaster: University of Lancaster PhD thesis, 1996), 383.

12 James A.W. Heffernan, "Ekphrasis and Representation," *New Literary History* 22, no. 2 (1991): 299. https://doi.org/10.2307/469040

13 Laura Baudot, "Air of History: Joseph Wright's and Robert Boyle's Air Pump Narratives," *Eighteenth-Century Studies* 46, no. 1 (2012): 1–28. https://doi.org/10.1353/ecs.2012.0075

14 Baudot, "Air of History," 8.

15 Heffernan, "Ekphrasis," 302.

16 Renate Brosch, "Ekphrasis in the Digital Age: Responses to Image," *Poetics Today* 39, no. 2 (2018): 227. https://doi.org/10.1215/03335372-4324420 See also Liliane Louvel, "Types of Ekphrasis: An Attempt at Classification," *Poetics Today* 39, no. 2 (2018): 245–63. https://doi.org/10.1215/03335372-4324432 , and Emma Kafalenos, "Ekphrasis as Misrepresentation: From Balzac's *Sarrasine* to Cortázar's 'Graffiti,'" *Poetics Today* 39, no. 2 (2018): 287–97. https://doi.org/10.1215/03335372-4324456

17 Robert Boyle, *New Experiments Physico-Mechanical, Touching the Spring of the Air, and its Effects* [1660], vol. 3, ed. M. Hunter and E.B. Davis (London: Pickering & Chatto, 2018), 274.

18 Boyle, *New Experiments*, 274.

19   Boyle, *New Experiments*, 274.

20   J.J. MacIntosh, "Animals, Morality and Robert Boyle," *Dialogue* 35 (1996): 457. https://doi.org/10.1017/S0012217300008817

21   Baudot, "Air of History," 15.

22   Steven Shapin and Simon Schaffer, *Leviathan and the Air Pump: Hobbes, Boyle, and the Experimental Life* (Princeton, NJ: Princeton University Press, 1985), 46.

23   Baudot, "Air of History," 15.

24   Robert Boyle, "New Pneumatical Experiments about Respiration," [1670], *The Works of Robert Boyle*, vol. 6, ed. M. Hunter and E.B. Davis (London: Pickering & Chatto, 2018), 221.

25   Joseph Conrad, *The N----- of the Narcissus*, ed. Allan H. Simmons (Cambridge: Cambridge University Press, 2017), 7.

26   Joseph Priestley, "An Account of Further Discoveries in Air," *Philosophical Transactions* 65 (1775): 384–94. https://doi.org/10.1098/rstl.1775.0039

27   Steven Connor, *The Matter of Air* (London: Reaktion Books, 2010), 26.

28   Priestley, "Further Discoveries," 388.

29   Daisy Lafarge, *understudies for air* (Bristol: Sad Press, 2017).

30   Lafarge, *Life*, 31.

31   Lafarge, *Life*, 31.

32   Lafarge, *Life*, 31.

33   Lafarge, *Life*, 36.

34   Lafarge, *Life*, 32, 8, 85.

35   Daniel Tiffany, "Lyric Substance: On Riddles, Materialism, and Poetic Obscurity," *Critical Inquiry* 28, no. 1 (2001): 75. https://doi.org/10.1086/449033

36   Lafarge, *Life*, 87.

37   Lafarge, *Life*, cover matter.

38   Jonathan Culler, *Theory of the Lyric* (Cambridge, MA: Harvard University Press, 2015), 35. https://doi.org/10.4159/9780674425781

39   Lafarge, *Life*, 83; Pasteur, *Studies*, 314.

40   Lafarge, *Life*, 83; Pasteur, *Studies*, 293.

41   Lafarge, *Life*, 85; Pasteur, *Studies*, 303.

42   Bruno Latour, "Pasteur on Lactic Acid Yeast: A Partial Semiotic Analysis," *Configurations* 1, no. 1 (1993), 129–46. https://doi.org/10.1353/con.1993.0004

43   Latour, "Pasteur," 132.

44   Latour, "Pasteur," 132.

45   Latour, "Pasteur," 140.

46   Lafarge, *Life*, 85.

47   Lafarge, *Paul* (London: Granta, 2021).

48   Chiang, "Exhalation," 37.

49   Chiang, "Exhalation," 37.

50   Chiang, "Exhalation," 48.

51   Chiang, "Exhalation," 50.

52   Chiang, "Story Notes," *Exhalation*, 342.

53   Chiang, "Exhalation," 50.

54   Peter Sloterdijk, *Terror from the Air* (Cambridge, MA: MIT Press, 2009), 22. See Siobhan Carroll's chapter in this volume, which raises an important challenge to the centrality of Ypres in Sloterdijk's history of this displacement.

55 Sloterdijk, *Terror*, 23.
56 Jean-Thomas Tremblay, "Homeostasis and Extinction: Ted Chiang's 'Exhalation,'" *SubStance: A Review of Theory and Literary Criticism* 52, no. 1 (2023): 22.
57 Tremblay, "Homeostasis," 25.
58 Tremblay, "Homeostasis," 22.
59 Gary K. Wolfe, "Alternate Cosmologies in American Science Fiction," *Hungarian Journal of English and American Studies* 18, no. 2 (2012): 170–71.
60 Wolfe, "Cosmologies," 171.
61 Tremblay, "Homeostasis," 27.
62 Tremblay, "Homeostasis," 27.
63 Peter Sloterdijk, *Spheres 1: Bubbles* (Cambridge, MA: MIT Press, 2011), 47.
64 Chiang, "Exhalation," 51.
65 Wolfe, "Cosmologies," 171.
66 Tremblay, "Homeostasis," 27-28.
67 Connor, *Matter*, 107, 337.
68 Connor, *Matter*, 337.
69 Connor, "Next to Nothing: The Arts of Air," *Art Basel*, 13 June 2007, http://stevenconnor.com/airart.html, accessed May 8, 2023.
70 Connor, "Next to Nothing."

8

The Importance of "Open Air" for Health: Environmental and Medical Intersections

Clare Hickman

In 1908, *The Medical Officer*, a professional journal aimed at those working as Medical Officers of Health in Britain, included a poster that had originally been created by the New York State Department of Health (see Fig. 8.1). The editors of the journal had sought permission for the poster to be reproduced in the UK for, as they note, "distribution in the homes of consumptives, ... posting in public places, such as the notice boards of churches, chapels &c."[1] This poster made a key visual link in the first line, where a series of pictograms outlined how "consumptives" (people suffering from tuberculosis) looking for a cure should consult the doctor, spend time in sunlight, get outdoor air, and rest.

The image chosen to accompany the outdoor air instruction is of a person with a hiking pole walking in hilly countryside, with a couple of pine trees depicted to the left of the image that presumably represent a forest.[2] The fact that someone hiking in the outdoors is depicted in medical promotional literature is key to this chapter. Rather than separating the medical from the environmental, it takes both a medical and environmental history approach to considering the use and promotion of the countryside in relation to health, including the late Victorian rise in campaigns for public access and retention of green spaces, such as commons, footpaths, forests, and parks. These arose alongside the adoption of open-air treatment in many institutions, particularly sanatoria for tubercular patients and open-air schools, and there are clear intersections between a medical practice that often also involved walking as part of the therapeutic regimen and the shared material culture of tents and chalets, which were used both medically as well as for leisure.[3] In particular, the environment and its perceived air quality is the focus of this chapter,

Clare Hickman, "The Importance of 'Open Air' for Health: Environmental and Medical Intersections" in: *Imagining Air: Cultural Axiology and the Politics of Invisibility*. University of Exeter Press (2023). © Clare Hickman. DOI: 10.47788/ERNO3036

Fig. 8.1. Detail from New York State Department of Health poster as reproduced in *The Medical Officer* on December 5, 1908. Credit: Wellcome Collection.

rather than the functions of air within the body itself, in relation to health and disease.[4] This extends and nuances the framework outlined by Christopher Sellers, specifically his argument that "both environmental and medical historians can seek to understand the past two centuries of medical history in terms of a seesaw dialogue over the ways and means by which physicians and other health professionals did, and also did not, consider the influence of place—airs and waters included—on disease."[5]

Ina Zweiniger-Bargielowska and many others have already argued that the concept of fitness was central to the late Victorian public health debate following the Boer War. As she notes, "late Victorian cultural pessimism about modernity, heightened by a growing awareness of international economic competition and imperial rivalry, resulted in extensive anxiety about physical deterioration or degeneration."[6] Similarly, David Matless has focused on ideas of national fitness in relation to physical uses of the countryside such as hiking, and Colin Ward and Dennis Hardy have developed analogous ideas in relation to leisure societies and holiday camps.[7] However, I want to go beyond this framing to consider in detail the role that the wider environment, and particularly medical concepts of the air in relation to particular places, fed into ideas regarding access to the countryside for leisure. In many ways, the dualism of urban versus rural to be explored here is found within the definition of the countryside itself. For example, the Cambridge Dictionary defines the countryside as "land not in towns, cities, or industrial areas, that is either used for farming or left in its natural condition."[8] This flattens the rural into something that encompasses everything from mountains to lowland agricultural land. The imagined nature of the healthy air of "countryside" or the "rural" can be read as oppositional to the nature of the unhealthy air of more urban spaces, and these broad terms have also influenced the ways in which medical practice has developed in relation to environmental health concerns.

This chapter concludes with an examination of the COVID-19 pandemic and the resurgence of ideas of open air in relation to health and disease prevention. The particularities of the medical knowledge related to disease transmission and prevention and the differences in the diseases themselves mean that this is very different in nature to the conception of open air a century ago. However, in practice these have all resulted in encouraging an increased use and value of the countryside and other green spaces for a range of activities.

The Therapeutic Concept of Open Air

In the description that accompanied the reproduction of the poster in *The Medical Officer* in 1908, it was noted that the health commissioner, Dr. Eugene Porter, who had shared the picture with the journal was also moving the Tuberculosis Exhibition to New York; this had originally been shown at the International Congress on Tuberculosis held in Washington in 1908. This exhibition was described as comprising "maps, charts, photographs, models of tents and sanatoria and every device known to modern science for use as a contributing agency in the prevention and treatment of tuberculosis. The exhibition halls have been partitioned off to represent town and country life, mountain and seashore."[9] Here, the environmental and topographical aspects of disease prevention as well as the apparatus of the sanatoria were clearly defined and understood by medical practitioners and shared via public health posters with the public. It is also clear that the open-air principles were reflected in the material culture, practices, and language applied across a range of different medical and leisure usages, and these will be explored further here.

The connections between air, the environment, and disease in relation to the Western medical tradition has ancient roots. As Jim Hankinson has observed in relation to the writers of the Hippocratic corpus (written in around the fourth century BCE): "pollutions are for them, as they still are for us, physical facts about the nature of the air and water themselves. The author of *Airs, Waters, Places* thus distinguishes between endemic and epidemic diseases, the former being a function of the site of the city in question."[10] These ideas about environmental factors are persistent and have been refor-mulated as concepts of disease causation and prevention change. For instance, when cholera epidemics started to occur in Britain in the 1830s, this was blamed on the "miasmas" or poisonous vapors in the air, which were consid-ered to be prevalent in poorer urban areas.[11]

There is similarly a long history of particular places being considered sinks of disease, accompanied by an idea that a change of air, or travel to a healthier location, would be the cure. In the early modern period, this could be achieved by those who could afford it by moving to a different part of the city, as Frances Gage has demonstrated in relation to Rome, or the countryside.[12] By the time of the cholera epidemics, the rapid program of urbanization and industrialization in Britain meant that one clear preventative approach to disease was time spent in less noxious air—which by this point was clearly signified by the countryside.

Spurred on by the public health concerns created by the first cholera epidemic of 1832, a House of Commons Public Health debate was held on February 21, 1833. During this, Mr. Slaney (MP for Shrewsbury) "rose for the purpose of moving that a Select Committee be appointed to consider the best means of securing open places in the neighbourhood of great towns, for the healthful exercise of the population."[13] Here the concerns stretched to the lower classes, who were living in cramped accommodation in narrow streets, and his proposal was to create public walks that would "afford exercise and recreation to the people."[14] Not everyone agreed, though. In the same debate, Mr. Baldwin, MP for Cork, argued that "cleanliness and the ventilation of houses, which were too much neglected in large towns, were more essential than public walks."[15]

In all cases, the debate revolved around ways of providing better air as well as offering an alternative to public houses (drinking establishments licensed to serve alcohol), which were seen as highly problematic for the working classes.[16] Mr, Slaney in his argument for public walks stated that "at present, the poor workman in the large manufacturing towns, was actually forced into the public house, there being no other place for him to amuse himself in."[17] It is also clear from these early debates that there were already local movements to maintain footpaths so that people could access the countryside as part of these wider health concerns. Slaney reports on a society in Manchester that had been set up to reopen footpaths that were being closed, stating that "artizans there had no place to walk on Sundays, and near that town and other such places, the working man and his family were met on the road with notices against trespass, and the inhospitable intimation of spring-guns, and Steel-traps."[18] Mr. Portman, MP for Marylebone, continued this line of argument, stating "that, in his opinion, the stopping up of foot-paths was one of the greatest causes of the want of sufficient opportunities of air and exercise for the humbler classes, and he hoped the House would not allow any opportunity to pass of doing all that legislation could do to prevent any further extension of that evil."[19] Here we already have connections between access to the countryside, epidemics, and public health that were developed and expanded upon as the air became seen as curative in itself, rather than an agent for causing or preventing disease.

Cholera outbreaks appeared in waves until the 1860s, but confirmation of its waterborne status changed medical understanding of its relationship to the air. Tuberculosis, however, was a more persistent disease with a clear connection to the lungs, and thereby air took on a particular importance by the late nineteenth century; at the same time, investments in public health infrastructures were beginning to reduce mortality from waterborne diseases. It is estimated that in 1850 tuberculosis was responsible for one in four deaths worldwide and that "until the 1870s it was the number one killer of Britons."[20] Given the scale, it is not surprising that ideas for prevention and treatment of tuberculosis loomed large in medical literature. By the end of the nineteenth century, even though rates were declining, the open-air approach, which in its most basic terms meant patients spending as much time as possible outside in fresh air and sunshine, became prevalent for many diseases, but most commonly tuberculosis. As F.B. Smith has noted, the vogue for open-air sanatoria began in Germany in around 1860, "inspired by a mixture of traditional cure-taking at spas, nature worship, and a new physiology of lung weakness."[21]

From this combination of factors, many doctors started developing approaches that emphasized the importance of the environment as a thera-peutic agent, which was also in part developed in response to a new understanding of natural immunity. According to Paul Weindling, these doctors "emphasised the importance of strengthening the body's natural resistance to disease rather than promoting 'artificial immunity' through mass immunisation."[22] This theory was summarized by Allan Warner, originally in a paper given at Leicester on "Open-air recovery schools" and then repub-lished in *The Medical Officer* in 1909, in which he argued that open-air therapies were gaining popularity in Britain. He argued that the "fresh air cure" had become to some extent a household word, but his belief was that, despite misconceptions of its role as a general panacea, it instead increased "metabolism and thus increase[d] the resisting power of the individual to the tubercle bacillus."[23] As Roger Cooter has argued, while the new bacte-riology theories of the 1870s still permitted air to be thought of as a reservoir of harmful agents, as in earlier theories of disease caused by miasma, they also allowed more emphasis to be placed on air as curative or constitutionally restorative in and of itself.[24]

These concepts formed the basis of the movement to place medical insti-tutions, such as sanatoria, in those rural places where the most curative air could be found, such as the Swiss Alps. In 1842, Dr. John Davy posited that a mountain climate was preferable because "alpine people rarely had tubercles because of the greater respiratory activity occasioned by a rarefied atmos-phere."[25] As Eva Eylers has argued in relation to Germany, where this idea was popularized, "the medical theory of the 'immune place,' developed by Hermann Brehmer in the 1850s, would serve as the impetus for the devel-opment and justification of the tuberculosis sanatorium, which, as the 'place of health,' was to be situated in natural surroundings, ideally in the dry air of an unspoilt mountain region."[26]

These themes can also be seen in the writings of those running sanatoria and other open-air institutions in Britain who were trying to compete with Alpine and German institutions and had to argue for the benefits of their own topographical locations in order to attract patients. Linda Bryder has argued that generally "British tuberculosis specialists held that the advantages of special climates had been greatly exaggerated and there was no therapeutic advantage in travelling to the Swiss Alps or the south of Europe," which is no surprise given the medical marketplace.[27] Richard Morris, in his work on travelling for health in this period, argues that "climate therapists disagreed—often heatedly—on which climates were best suited to treating consumption."[28] However, local climatic conditions were generally noted in advertisements, with the Vale of Clwyd Sanatorium, for example, stating that, "it has a bracing climate, with a small rainfall and a large amount of sunshine."[29] Similarly, particular coastal places, such as Bournemouth, also developed a reputation as a health resort owing to their particular climatic conditions.[30] F.W. Burton-Fanning, a key proponent of open-air institutions in Britain, noted that "we like the treatment to be carried out in a locality whose air is pure and bracing, and moderately dry."[31] The height of the institution above sea-level appears to have been important in many advertisements, which could also relate to the supposed efficacy of higher mountainous locations.

One key example that will be used here to unpick some of these connections will be the writings of David Chowry Muthu, an Indian physician who established the Mendip Hills Sanatorium in the UK and the Tambaram Sanatorium in India. He wrote several books on his experience of running an open-air sanatorium in the Mendips, and his works offer a lens on to his use of the rural environment as part of his therapeutic approach. For instance, in 1910, Muthu argued that open-air therapy was a "natural treatment" and suggested that, "the secret of its widespread interest in Europe is due to the discovery—if discovery it may be called—that fresh air, hitherto regarded as an enemy to be shut out and barred, is really a friend, and one of Nature's best gifts to man." [32] Like many, he related the concept of open-air therapies to a time before industrialization and urbanization, "that man, by building towns and manufacturing dirt and disease, is undoing Nature's work, and that to put himself right again he must go back to Nature, and lead an open-life in the green fields and meadows, and breathe the sweet fresh air."[33]

At the same time, Lord Eversley, in his work *The Story of the Battle during the Last Forty-five Years for Public Rights over the Commons, Forests and Footpaths of England and Wales*, noted the importance of access to country air for urban dwellers:

> Where such Metropolitan or Suburban Commons exist it is difficult to exaggerate their value to the public. They are natural parks, over which everyone may roam freely; for though the public may be trespassers in strict law, there are no practical means of preventing the use of these waste lands for exercise and recreation. They are reservoirs of fresh air and health, whence fresh breezes blow into the adjoining town. They bring home to the poorest something of the sense and beauty of nature.[34]

Here, the air from commons was perceived as refreshing the city as well as commons themselves giving places for people to roam, and there are clear connections between his arguments about the power of nature and the need for urban dwellers to access fresh, rural air for health to those of Muthu with his conception of nature as a powerful therapeutic force.[35]

This reflects what Bill Luckin and Keir Waddington have described as the pro-rural/anti-urban sentiment that emerged in the nineteenth century and was strengthened by the concerns around physical and mental degeneration of the population at the end of the nineteenth century.[36] As Zweiniger-Bargielowska notes, there were fears that "the capital's population was degenerating into 'a puny race unfit to maintain themselves' due to the foul air and artificial, morally corrupting living conditions of the modern city."[37] One of the examples she cites is that of James Cantlie, assistant surgeon at Charing Cross Hospital, who argued that it is "'difficult—I will not say impossible—to find a third generation of pure Londoners' because 'children of these late generations seldom reached maturity.'"[38] Air and its surrounding environment were believed by both the general public and medical practitioners to be key elements in relation to the promotion of health or causation of disease. All of this ignored actual health problems experienced by those in the countryside, including the agricultural recession of the 1870s–1890s.[39]

The work by Waddington on the perception of water in rural Wales acts as a useful correction and highlights the issues of a shared cultural and medical imagining of the countryside as a place of health.[40] For instance, although Wales in the late nineteenth century may have been "widely perceived as a wet and wild place"—a view reinforced by Romantic paintings and guidebooks—there was often a scarcity of usable water for people in rural areas.[41] As Waddington notes, "if water from isolated and mountainous regions could be imagined as examples of 'perfect purity,' investigations found that many local wells and springs in rural Wales were 'not in a satisfactory state' and often 'tainted.'"[42] The commonly shared understanding of the nature of particular environments, in this case Wales, could lead to medical professionals and others ignoring actual health issues related to those resources and scarcity of supply for human populations, or overlooking other problems such as contamination owing to the conception of clean mountain spring water. In the same way, the imaginations of clean air helped create the concept of the countryside as a healthy space. It was this perception, rather than the actuality of countryside or rural areas, that dominated the cultural context and shaped both medical practice and wider perception of countryside as the imagined antithesis of a polluted and degenerate environment.

Similarly, medical professionals in some cases also appear to have viewed the garden city, which blended the countryside with the town, as a potential weapon in the war against the degeneration of the population. In 1919, a writer in *The Medical Officer* argued that garden cities could help create a more vigorous and happy nation as it was:

> [O]pen air and exercise, good feeding and well-regulated rest which had converted weedy citizens into robust soldiers, which restored the weakly in open-air schools, and the consumptive in sanatoria. ... If for that reason alone, Dr Hill, urged that garden cities should be built and big towns not added to.[43]

Again, we can see the connections between the open-air movement and the imagined ideal of the healthy countryside combined in a bid to improve the health of future generations through both access to a healthier environment and through physical fitness.

Material Cultures and Practices

By analyzing the material structures created for patient use in relation to open-air therapies, as well as the wider concept of air as curative, we can see clear links with the emerging holiday movement; specifically the use of sleeping chalets and tents. At Hailey Open Air Sanatorium, near Wallingford, Oxfordshire (opened c.1890s) there are photos depicting sleeping chalets in the grounds and a tent in the meadow (see Fig. 8.2). These were crucial elements for the implementation of the open-air method being practiced at Hailey in the same manner as Muthu and others. According to Charles Reinhardt, the medical superintendent, "Hailey is, with the exception of Alderney Manor, the only British Sanatorium built entirely on the separate chalet or hut principle."[44] The hut or tent was the ideal structure in which to obtain the greatest benefits from fresh air. As one medical practitioner argued, "all that a consumptive requires in the shape of accommodation consists of a hut, which must have windows opening in every direction. In such a contrivance he can breathe air identical with that of the open country, and in no room of a building is this possible."[45] Again, access to a maximum amount of "fresh" country air was the key thing, and huts were seen as being able to facilitate this in a way that buildings could not.

Similarly, verandas were constructed so that patients could sit out of the wind and balconies were created so that patients could be wheeled out on to them without leaving their beds to get some benefit from the fresh air and sunshine (see Fig. 8.3). Margaret Campbell has quite rightly argued that these can be seen as features relating to the summer house, gazebo, or shelter that were familiar sights in eighteenth-century landscapes, and that they were, as well as being used by institutions, common features of domestic gardens for the use of a sick family member.[46] The choice of designs available reflects their inclusion as part of an aesthetic landscape as well as a medical one. There was a competitive market for chalets and huts for domestic use for those who chose to be treated at home, as can be seen in advertisements of the time.[47]

Other influences on domestic architecture included the use of balconies and the design of reclining chairs, such as those based on the recliners designed by Alvar Aalto for the Paimio Sanatorium; although, as Campbell makes clear, it cannot be argued "that the introduction of the flat roof, balcony, summer house and recliner chair were the direct result of early treatment methods for tuberculosis." However, "the popularity of these modernist archi-tectural features in the pursuit of good health and hygiene placed them in the annals of a therapeutic lifestyle," one that, of course, has since been abandoned owing to new therapies such as antibiotics and other developments such as vaccines.[48]

A SLEEPING CHÂLET WITH VERANDAH, HAILEY SANATORIUM,
WALLINGFORD.

Fig. 8.2. A sleeping chalet with veranda at Hailey Sanatorium, Wallingford, featured in
Charles Reinhardt, *A Handbook of the Open-Air Treatment and Life in an Open-Air
Sanatorium* (London: John Bale, Sons & Danielson, 1902). Credit: Wellcome
Collection.

Fig. 8.3. Winford Orthopaedic Hospital, Bristol with patients wheeled out to receive fresh air and sunlight. From the *Winford Orthopaedic Hospital Annual Report*, 1936. Credit: With the permission of the University of Bristol Special Collections.

The popularity of chalets, tents, and time spent in the countryside for leisure arose at the same time and were surely in part influenced by wider public understanding of health in relation to the open-air movement.[49] As Morris has demonstrated, reasons for travelling for health and/or pleasure are complex and intertwined in their relationship to cultural and social under-standing of diseases, and this approach can be applied to these wider developments of holiday camps.[50] For the Woodcraft Folk, a youth organization where camping is still a key element, there were clear intersections with the use of the outdoors in sanatoria and other institutions. In the 1930s, they stated that "camp life combines the general romance of the movement with the material delight of sleeping in tents, of making fires, bathing in sun and water, tramping over the hills and singing round the fire."[51] Here the same practices were used with a similar aim: "to provide them [young people] with opportunities for healthy outdoor activity."[52] As late as 1944, the text on a membership card for the Darlington Cooperative Holidays Association and Holiday Fellowship Joint Group was still stating that their "Objects" were "to provide for the healthy enjoyment of leisure" and to "encourage a love of the open air."[53] So the wider public understanding of open-air for health was able to move from the practices of the sanatoria. This extends Graham Mooney's argument that "the domestication of tuberculosis was achieved by moulding elements of existing public health policy with components of the sanatorium regimen into a kind of domesticated 'preventive therapy' for tuberculosis."[54] Here, the domestic moved beyond the house and into the wider use of the countryside for leisure, but there was a similar blending and adaptation of the ideals of the therapeutic regimen into daily life.

Tramping

Similarly, Muthu's approach also saw "tramping" or long-distance walking as a key element in the full recovery of his patients. As part of his therapeutic regime, he included exercise under the following five headings that tracked patient recovery: "(1) walking exercises; (2) graduated exercises; (3) breathing exercises; (4) singing exercises; (5) camping, tramping etc."[55] Patients who were lucky enough to keep improving in health would be encouraged once they were fit enough to travel even further into the countryside on tramping tours as part of their recovery (see Fig. 8.4). He wrote:

> By the time the patients reach the convalescent period they instinc-
> tively desire a wider scope and greater activity to give exercise to
> their returning health. Therefore, we saw the necessity of introducing
> a system of tramping tours during the last stage of the patients' stay
> in the sanatorium as a means of giving a final touch to the cure of
> their disease. And the result has been eminently satisfactory. By
> taking the patients to "fresh fields and pastures new," a new interest
> has been created which gave fresh stimulus to the healing energies
> to bring about the final arrest of the disease.[56]

These excursions were built up over time from half-day walks in the afternoon covering about 5 to 6 miles to much longer walks of 12 miles over a whole day. Once patients were easily covering those miles, these walks would ideally lead to "caravan tours," with Muthu writing that the "patients on these tours go still further afield, and after tramping for three or four days in the open country and living a simple life, return to the sanatorium."[57] This reflected the growing interest in walking-based holidays to the countryside, and combined camping as a healthy as well as a pleasurable practice.

Muthu clearly felt that there was a preventive health element to this form of sanatorium practice for the wider public as well as benefits for convalescent patients. He argued more broadly that benefits would accrue to many people living in towns from this kind of activity:

> What a change this tramping life would bring to the town-bred
> patient. The rosy flush of the early dawn, the morning chant of the
> birds, the sunlit fields and meadows fill him with new delights; while
> the afterglow of the sunset, the hush of the twilight, the radiance of
> the starlit sky calm his mind and still his soul into quietness and
> peace. And as he lies down to rest he nestles close to the bosom of
> mother Nature and lets her wrap him in a dreamless sleep.[58]

This is a particularly idealized description of the countryside, but one that is common in the literature describing the benefits of the countryside. As Hester Barron notes, "the interwar years saw the rural ideal raised by some

FIG. 23.—PATIENTS ON A TRAMPING TOUR.

To face page 246.

Fig. 8.4. Photograph of patients on a tramping tour from Muthu, *Pulmonary Tuberculosis: Its Etiology and Treatment a Record of Twenty Two Years' Observation and Work in Open-Air Sanatoria* (London: Balliere, Tindall and Cox, 1922). Credit: Wellcome Collection.

to an almost spiritual level," which is reflected in Muthu's highly Romantic account of the sensory experience of walking and sleeping in the countryside.[59] This is not incidental. As Bharat Jayram Venkat has stated, as well as training with well-known physicians in Britain, Muthu also joined "the YMCA and the British temperance movement, was nicknamed 'the Christian Brahmin' by the papers," and married a well-connected woman whose family belonged to the gentry, so was very much part of the Christian establishment espousing these claims of the countryside as a place that was physically and morally uplifting.[60] However, in his own work he also talked about Hindu approaches, writing that "perhaps the Hindu philosophers are right after all when they assert that besides its chemical constituents the atmospheric air contains a vital universal principle called 'prana,' through which life manifests itself, and the fresher the air, the more it is charged with 'prana.'"[61] This suggests a different spiritual underpinning but one that also saw fresh air as crucial to a sound physical body.

These healthy and morally uplifting ideals, often promoted by the middle and upper classes for the stated benefit of the working classes, also influenced the rise of campaigning groups including the Council for the Preservation of Rural England (1926), the Youth Hostel Association (1930), and the Ramblers' Association (1935), as well as youth organizations such as the Woodcraft Folk (1925), mentioned earlier.[62] As Barron argues, these all became popular as "scientific and medical emphasis on the restorative effect of fresh air became ubiquitous in interwar Britain and Europe."[63] There was then a clear influence from the wider medical perceptions of the role of air and sun in health on these organizations and the broader campaign for access to the countryside. As Bryder has noted, the National Association of Prevention of Tuberculosis's first sound film, called *Stand up and Breathe* and made in 1935, focused on health and included images of hikers with their dogs, highlighting the strength of the interconnections.[64]

The spiritual was also a wider element in discourses around physical activity in rural areas. As Mark Freeman has stated, there were strong spiritual as well as political dimensions to much of the early movement involved in activities such as long-distance walking and mountaineering.[65] Rambling, for example, was already popular in parts of Britain by the end of the late nineteenth century: Some regions such as northeast Lancashire had established clubs that were organized via Nonconformist churches.[66] Freeman makes similar points about the emergent Quaker tramp movement, where the "simple open-air life" was one of the elements that they felt brought "the Kingdom of God nearer."[67] These ideals of spiritual renewal in the countryside as opposed to urban areas had already been summarized by William Cowper in 1785 as "God made the country, and man made the town."[68]

Beauty and quiet, as well as fresh air, became increasingly invoked in late Victorian and Edwardian texts arguing for the beneficial nature for body and spirit of spending time in gardens and rural places. In 1883, Octavia Hill, one of the founders of the National Trust, argued that "we all need space; unless we have it we cannot reach that sense of quiet in which whispers of better things come to us gently. Our lives in London are over-crowded, over-excited, over-strained. This is true of all classes; we all want quiet; we all want beauty for the refreshment of our souls."[69] She continued that "sometimes we think of it as a luxury, but when God made the world, He made it very beautiful, and meant that we should live amongst its beauties, and that they should speak peace to us in our daily lives."[70]

As highlighted by Amelia Bonea et al., popular medical works, such as Dr. Benjamin Ward Richardson's *Diseases of Modern Life* (1876), talked about "'diseases from worry and mental strain,' and also illnesses generated by physical conditions such as poisoned air, as well as what we would now call 'lifestyle diseases.'"[71] As the authors argue, there were clear concerns about rapid changes in how bodies interacted with, and were situated within, rapidly changing social and physical environments.[72] Similarly, according to Stephen Arata, there were "innumerable late-Victorian accounts of the malady of modern life make their way round to the issue of sensory overload: too many images, too much noise, way too much information, all of it too often resulting in nervous collapse, neurosis, dysfunctions of various kinds."[73]

The call to "quiet" may have lost some of its religious connotations, but as Robert Snape has noted, The Edwards Report of 1991 ("Fit for the Future: Report of the National Parks Review Panel"), "re-affirmed this approach in advocating not only the facilitation but the promotion of the quiet enjoyment of the countryside," and that more recent work has demonstrated that "concepts of peace and quiet have become dominant aspects of strategic management in National Parks."[74] To mark its 125th anniversary in 2020, the National Trust also used Octavia Hill's words—"We all want quiet. We all want beauty. We all need space"—as key mottos for their present-day work, which reflects current fears regarding technology and modernity in relation to our own health today.[75] Robert Louv's 2005 nature deficit disorder concept, which has being popularly taken up as a way to understand what are perceived to be contemporary health problems caused by technology overload, particularly for children, can be seen to have strong resonances with these ideas of the past where rural restoration is seen as both preventative and curative.[76]

COVID-19 and Open Air

As the discussion here has demonstrated, there are particular medical under-pinnings that affect the way that air and its intimate connections to an imagined countryside has been perceived in relation to health and disease, although the general sense that there are inequalities in who can access healthy (often rural) air that need to be resolved have remained the same. COVID-19 has revealed similar patterns of concern. In statements about access to green spaces during the pandemic, there are clear echoes to medicalized views of the countryside that emerged after the cholera epidemics and calls for public walks so people can access cleaner air—even though the understanding of the disease, its prevention, and its cure are medically very different. A strong reflection of earlier arguments about the importance of fresh rather than infected city air were evident in statements made, such as this one by Cheshire East council during the COVID-19 pandemic in the UK:

> Public rights of way, country parks and trails in the countryside and urban fringe provide an opportunity for people to take exercise and get some fresh air in these difficult times. The government reports that the risk of the coronavirus being passed on to others from people using outdoor green spaces, paths and trails is considered to be very low.[77]

The second part of this statement reflects the contemporary understanding of the role of air in disease transmission, but the first part on the importance of fresh air invokes older understandings of the role of country air in health. Like Lord Eversley in 1910, the green fringes and rights of way are perceived as offering opportunities for healthy activity in cleaner air. As the UK-wide People and Nature Survey run by Natural England throughout the pandemic noted in its summary of findings, "Nearly half of adults in England report spending more time outdoors than before the pandemic (45%, March 2022). Close to four in ten say that nature and wildlife are more important than ever to their wellbeing (39%, March 2022)."[78] There is of course always a specific medical context, as this chapter has tried to demonstrate, but the general argument for more access to these "healthier" outdoor spaces during times of epidemic disease has long roots.

Similarly, these moments of public health concern have also highlighted inequalities in who can access these "healthier" locations. Previously, this has generally referred to lower-class populations, and as we have seen particularly children, but with population changes this has shifted slightly in focus. In the 2020 report from the Ramblers, "The Grass Isn't Greener for Everyone: Why Access to Green Space Matters," they noted:

> [O]nly 57% of GB adults questioned said that they lived within five minutes' walk of green space, be it a local park, nearby field or canal path. That figure fell to just 39% for people from a Black, Asian or Minority Ethnic (BAME) background and 46% among all GB adults with a household income of under £15,000 (compared to 63% of those with a household income over £35,000 and 70% over £70,000).[79]

Reflecting the same paternalistic concerns of the nineteenth-century access to the countryside movement, there is again a keen understanding of inequalities between those who can easily visit these spaces and who is excluded. As the 2021 British Academy report "The COVID Decade: Understanding the Long-Term Societal Impacts of COVID-19" noted, "throughout the pandemic, access to green spaces has proven beneficial to health and wellbeing—but this need has further highlighted social disadvantage."[80]

As in the past, though, attention needs to be paid to shared narratives that highlight the healthy nature of rural air but might cause other data to be overlooked. As the 2020 Nuffield report into rural health services and their potential recovery post-COVID-19 noted, "history suggests that rural and remote services are at risk of not getting what might be deemed their fair share of the additional funding that is being injected into the NHS as a result of the pandemic."[81] There are many and varied reasons for this, but an overarching conception of the rural as healthy might cause policymakers to assume that cities have more need of funding. Similarly, just because rural places are in the countryside does not necessarily mean that green spaces are accessible. As Natural England stated in their 2010 report, access to high quality natural space is "often more achievable in urban communities than in rural communities, particularly in lowland agricultural England where there is often poor access to quality greenspace."[82] There is therefore a need for more attention to the ways in which imaginaries of clean, healthy air and the wider countryside might mask the complex social, cultural, and health issues faced by rural communities.

Conclusion

By focusing on the countryside, air, and medical concepts of health and disease during public health events over two centuries, the tensions between what is conceived as modern scientific knowledge and more nostalgic ideas of health are clear. As Christopher Hamlin has argued, "a curious anomaly of medical history is the representation of environmental health concerns simultaneously as very recent—as predicated on developments in the environmental sciences; and as very ancient, exotic, and outdated—an exemplar of a prescientific age."[83] In many ways, these ideas of both the Ancient and the modern Western medical mindset are not only seen in the work of historians, but they are also reflected in the way the environment, and particularly air, has been conceived in relation to disease by the medical profession at the time. The tuberculosis sanatorium, for instance, was simultaneously both very modern in its architectural design and understanding of bacteriology and premodern in its emphasis on the role place and environment could play. However, by focusing on the scientific understanding of open air as a therapeutic tool, we can also see that as well as looking backwards to the preindustrial age, air itself was also understood in relation to bacteriology and laboratory medicine.

Throughout these examples, there are contradictions between ideas of nature and modernity, but perhaps these ideas are best understood as entangled rather than in opposition, as nature is perhaps only really understood in relation to what is an urban, industrial society. As William Cronon argues, "the dream of an unworked natural landscape is very much the fantasy of people who have never themselves had to work the land to make a living."[84] The calls to leisure time in the countryside for health were similarly centered within sometimes contradictory ideas of nostalgia on the one hand and the cult of modern fitness on the other. By taking both an environmental and medical history lens to thinking about air in relation to disease causation and treatment, these complex entanglements can be brought to the surface, and we can gain a greater sense of the relationship between health, leisure, and medical practice.

Acknowledgments: This research was supported by the Arts and Humanities Research Council [AH/V00509X/1] as well as discussions developed as part of the Wellcome Trust funded, "MedEnv: Intersections in Medical and Environmental Humanities Network" [218165/Z/19/Z]. With thanks to Abbi Flint, Glen O'Hara and Keir Waddington for commenting on drafts of this work.

Notes

1   This poster was reproduced in *The Medical Officer* on December 5, 1908, 432, and May 3, 1909, 205.

2   I have explored the significance of pine trees in more detail here: "Pine Fresh: The Cultural and Medical Context of Pine Scent in Relation to Health—from the Forest to the Bathroom," *Medical Humanities* 48 (2022): 104–13. https://doi.org/10.1136/medhum-2020-012126

3   For more on open-air schools, see Clare Hickman, "Care in the Countryside: The Theory and Practice of Therapeutic Landscapes in the Early Twentieth-Century," in *Landscape and Green Spaces: Gardens and Garden History in the West Midlands*, ed. Malcolm Dick and Elaine Mitchell (Hatfield: Hertfordshire University Press, 2018), 160–85.

4   For more on the technological engagement with ideas of air and the body, see Jennifer Wallis, "A Machine in the Garden: The Compressed Air Bath and the Nineteenth-Century Health Resort," in *Histories of Technology, the Environment and Modern Britain*, ed. Jon Agar and Jacob Ward (London: UCL Press, 2018), 76–100. https://doi.org/10.2307/j.ctvqhsmr.9

5   Christopher Sellers, "To Place or Not to Place: Toward an Environmental History of Modern Medicine," *Bulletin of the History of Medicine* 92, no. 1 (2018): 6. https://doi.org/10.1353/bhm.2018.0000

6   Ina Zweiniger-Bargielowska, *Managing the Body Beauty, Health, and Fitness in Britain, 1880–1939* (Oxford: Oxford University Press, 2010), 6. https://doi.org/10.1093/acprof:oso/9780199280520.001.0001

7   David Matless, *Landscape and Englishness*, 2nd expanded ed. (London: Reaktion, 2016), 131–37. See also Colin Ward and Dennis Hardy, *Goodnight Campers!: The History of the British Holiday Camp* (Nottingham: Five Leaves Publications, 2010) on these wider health and leisure ideas.

8   Cambridge University Press, *Cambridge Dictionary Online*, https://dictionary.cambridge.org/dictionary/english/countryside, accessed February 9, 2023.

9   *Medical Officer*, December 5, 1908, 432.

10 Jim Hankinson, "Aetiology," in *The Cambridge Companion to Hippocrates*, ed. P. Pormann (Cambridge: Cambridge University Press, 2018), 113. https://doi.org/10.1017/9781107705784.006

11 Pamela K. Gilbert, "On Cholera in Nineteenth-Century England," in *BRANCH: Britain, Representation and Nineteenth-Century History*, ed. Dino Franco Felluga. Extension of *Romanticism and Victorianism on the Net*, https://branchcollective.org/?ps_articles=pamela-k-gilbert-on-cholera-in-nineteenth-century-england, accessed July 13, 2022.

12 Frances Gage, "Chasing 'Good Air' and Viewing Beautiful Perspectives," in *Conserving Health in Early Modern Culture*, ed. Sandra Cavallo and Tessa Storey (Manchester: Manchester University Press, 2017), 238.

13 House of Commons Debate, February 21, 1833, vol. 15, cc1049–59.

14 House of Commons Debate.

15 House of Commons Debate.

16 Clare Hickman, "An Exploration of the National Health Society and Its Influence on the Movement for Urban Green Spaces in Late-Nineteenth Century London,"

Landscape and Urban Planning 118 (2013): 112–19. https://doi.org/10.1016/j. landurbplan.2012.09.007

17   House of Commons Debate.
18   House of Commons Debate.
19   House of Commons Debate.
20   Richard E. Morris, "The Victorian 'Change of Air' as Medical and Social Construction," *Journal of Tourism History* 10, no. 1 (2018): 51.
21   F.B. Smith, *The Retreat of Tuberculosis, 1850–1950* (London: Croom Helm Ltd, 1988). 97.
22   Paul Weindling, "From Infectious to Chronic Disease: Changing Patterns of Sickness in the Nineteenth and Twentieth Centuries," in *Medicine in Society*, ed. Andrew Wear (Cambridge: Cambridge University Press, 1992), 313–14. https://doi.org/10.1080/1 755182X.2018.1425485
23   Allan Warner, originally a paper given at Leicester, "Open-Air Recovery Schools," republished in *The Medical Officer*, December 25, 1909, 469.
24   Roger Cooter, "Open-Air Therapy and the Rise of Open-Air Hospitals," in *The Bulletin for the Social History of Medicine* 35 (1984): 44.
25   As quoted in Susan Barton, *Healthy Living in the Alps: The Origins of Winter Tourism in Switzerland, 1860–1914* (Manchester: Manchester University Press, 2008), 10.
26   Eva Eylers, "Planning the Nation: The Sanatorium Movement in Germany," *The Journal of Architecture* 19, no. 5 (2014): 667. https://doi.org/10.1080/13602365.201 4.966587
27   Linda Bryder, *Below the Magic Mountain: A Social History of Tuberculosis in Twentieth-Century Britain* (Oxford: Clarendon Press, 1988), 46.
28   Morris, "The Victorian 'Change of Air,'" 53.
29   Advertisement in *The Medical Officer*, September 5, 1908, 41.
30   Hickman, "Pine Fresh."
31   F.W. Burton-Fanning, *Open-Air Treatment of Pulmonary Tuberculosis* (London, Paris, New York, and Melbourne: Cassell and Company, c.1900), 160.
32   David Chowry Muthu, *Pulmonary Tuberculosis and Sanatorium Treatment: A Record of 10 Years' Observation and Work in Open-Air Sanatoria* (London: Bailliere, Tindall & Cox, 1910), 83–84.
33   Muthu, *Pulmonary Tuberculosis and Sanatorium Treatment*.
34   George Shaw-Lefevre (Lord Eversley), *Commons, Forests and Footpaths. The Story of the Battle during the Last Forty-Five Years for Public Rights over the Commons, Forests and Footpaths of England and Wales*, revised edition [of *English Commons and Forests*] (London: Cassell and Company, 1910), 3.
35   For more on this see Karen Jones, "'The Lungs of the City': Green Space, Public Health and Bodily Metaphor in the Landscape of Urban Park History," *Environment and History* 24 (2018): 39–58. https://doi.org/10.3197/096734018X15137949591837
36   Bill Luckin, *Death and Survival in Urban Britain: Disease, Pollution and Environment, 1800–1950* (London: Bloomsbury, 2015) https://doi.org/10.5040/9780755621576; Keir Waddington, "'In a Country Every Way by Nature Favourable to Health': Landscape and Public Health in Victorian Rural Wales," *Canadian Bulletin of Medical History* 31, no. 2 (2014): 183–204. https://doi.org/10.3138/cbmh.31.2.183
37   Zweiniger-Bargielowska, *Managing the Body Beauty*, 24.
38   Zweiniger-Bargielowska, *Managing the Body Beauty*, 25.

39 Luckin, *Death and Survival in Urban Britain*, 145.

40 Keir Waddington, "'I should have thought that Wales was a wet part of the world': Drought, Rural Communities and Public Health, 1870–1914," *Social History of Medicine* 30, no. 3 (2017): 590–611. https://doi.org/10.1093/shm/hkw118

41 Waddington, "Drought, Rural Communities and Public Health," 595

42 Waddington, "Drought, Rural Communities and Public Health," 595.

43 *The Medical Officer*, July 12, 1919, 11.

44 Charles Reinhardt, *A Handbook of the Open-Air Treatment and Life in an Open-Air Sanatorium*, second edition (London: John Bales, Sons and Danielson, 1902), 123.

45 "The Financial Aspect of the Open Air Treatment," *Health Resort*, 1903, 149, quoted in Margaret Campbell, "What Tuberculosis Did for Modernism: The Influence of a Curative Environment on Modernist Design and Architecture," *Medical History* 49, no. 4 (October 2005): 479. https://doi.org/10.1017/S0025727300009169

46 Campbell, "What Tuberculosis Did for Modernism, 478–80.

47 Advertisements for a variety of different architectural styles for open-air huts were published in the *British Journal for Tuberculosis* including Japanese style buildings.

48 Campbell, "What Tuberculosis Did for Modernism, 488.

49 For more on this rise of tents and chalets for holidays and leisure see Ward and Hardy, *Goodnight Campers!*.

50 Morris, "The Victorian 'Change of Air.'"

51 Morris, "The Victorian 'Change of Air,'" 24.

52 Morris, "The Victorian 'Change of Air,'" 24.

53 Membership card for the Darlington Cooperative Holidays Association and Holiday Fellowship Joint Group, National Cooperative Archive, Manchester, CHA/3/1.

54 Graham Mooney, "The Material Consumptive: Domesticating the Tuberculosis Patient in Edwardian England," *Journal of Historical Geography* 42 (2013): 152–53. https://doi.org/10.1016/j.jhg.2012.12.007

55 David Chowry Muthu, *Pulmonary Tuberculosis: Its Etiology and Treatment a Record of Twenty Two Years' Observation and Work in Open-Air Sanatoria* (London: Balliere, Tindall and Cox, 1922), 236.

56 Muthu, *Pulmonary Tuberculosis*, 244.

57 Muthu, *Pulmonary Tuberculosis*, 246.

58 Muthu, *Pulmonary Tuberculosis*, 246.

59 Hester Barron, "Changing Conceptions of the 'Poor Child': The Children's Country Holiday Fund, 1918–1939," *The Journal of the History of Childhood and Youth* 9, no. 1 (2016): 32. https://doi.org/10.1353/hcy.2016.0018

60 Bharat Jayram Venkat, *At the Limits of Cure* (Durham, NC: Duke University Press, 2021), 36.

61 Muthu, *Pulmonary Tuberculosis*, 232.

62 Barron, "Changing Conceptions of the 'Poor Child,'" 32.

63 Barron, "Changing Conceptions of the 'Poor Child,'" 32.

64 Bryder, *Below the Magic Mountain*, 152–53.

65 Mark Freeman, "Fellowship, Service and the 'Spirit of Adventure': The Religious Society of Friends and the Outdoors Movement in Britain c.1900–1950," *Quaker Studies* 14, no. 1 (2009): 72–92. https://doi.org/10.3828/quaker.14.1.72

66 Robert Snape, "The Co-operative Holidays Association and the Cultural Formation of Countryside Leisure Practice," *Leisure Studies* 23, no. 2 (2004): 145.

67 Freeman, "Fellowship, Service and the 'Spirit of Adventure'": 77. https://doi. org/10.1080/0261436042000226345

68 William Cowper, *The Task and Other Poems* (London: Cassell and Company, 1896), 33.

69 Octavia Hill, *Homes of the London Poor* (London: Macmillan and Co., 1883), 95.

70 Hill, *Homes of the London Poor*, 95.

71 Amelia Bonea, Melissa Dickson, Sally Shuttleworth, and Jennifer Wallis, *Anxious Times: Medicine and Modernity in Nineteenth-Century Britain* (Pittsburgh, PA: University of Pittsburgh Press, 2019), 4. https://doi.org/10.2307/j.ctvk8w1tx

72 Bonea et al., *Anxious Times*, 5.

73 Stephen Arata, "On Not Paying Attention," *Victorian Studies* 46, no. 2 (2004): 198. https://doi.org/10.1353/vic.2004.0077

74 Snape, "The Co-operative Holidays Association," 144.

75 The National Trust film made for this highlights their use of Hill's words for contemporary audiences (https://www.youtube.com/watch?v=zcsLWqO0RL0, accessed May 9, 2023). Some of these themes are also explored by Agnes Arnold-Forster in "Still the Same: Nineteenth-Century Critiques of Technology Show How Longstanding Many Current Concerns Are," *Real Life*, June 27, 2022, https:// reallifemag.com/still-the-same/, accessed July 13, 2022.

76 Colin Tudge and Aleks Krotoski, "Are Modern British Children Suffering from 'Nature Deficit disorder'?," *The Guardian*, March 30, 2012.

77 Cheshire East Council, "Covid-19 and Countryside Sites," https://www.cheshireeast. gov.uk/leisure,_culture_and_tourism/ranger_service/covid-19-and-countryside-sites. aspx, accessed June 20, 2022.

78 Tom Marshall, "People and Nature Survey: How Has COVID-19 Changed the Way We Engage with Nature?," 18 May 2022, https://naturalengland.blog.gov. uk/2022/05/18/people-and-nature-survey-how-has-covid-19-changed-the-way-we-engage-with-nature/, accessed July 13, 2022.

79 Quote from this blog: https://www.ramblers.org.uk/news/blogs/2021/january/ securing-a-better-future-for-public-access-to-nature.aspx. The full report is Ramblers, "The Grass Isn't Greener for Everyone: Why Access to Green Space Matters," 2011, http://www.ramblers.org.uk/thegrassisntgreener, both accessed July 13, 2022.

80 *The COVID Decade: Understanding the Long-Term Societal Impacts of COVID-19* (London: The British Academy, 2021), 62.

81 William Palmer and Lucina Rolewicz, *Rural, Remote and at Risk: Why Rural Health Services Face a Steep Climb to Recovery from Covid-19* (London: Nuffield Trust, 2020), 8.

82 Natural England, *'Nature Nearby': Accessible Natural Greenspace Guidance* (2010), 9.

83 Christopher Hamlin, "Surgeon Reginald Orton and the Pathology of Deadly Air: The Contest for Context in Environmental Health," in *Toxic Airs: Body, Place, Planet in Historical Perspective*, ed. James Rodger Fleming and Ann Johnson (Pittsburgh, PA: University of Pittsburgh Press, 2014), 25–26.

84 William Cronon, "The Trouble with Wilderness; or, Getting Back to the Wrong Nature," in *Uncommon Ground: Rethinking the Human Place in Nature*, ed. William Cronon (New York: W. W. Norton & Co., 1995), 80.

9

"The Endless Space of Air": Helen Keller's Auratic Worldbuilding

Jayne Lewis

What then is the aura? A strange tissue of space and time: the unique apparition of a distance, however near it may be. To follow with the eye— while resting on a summer afternoon—a mountain range on the horizon or a branch that casts its shadow on the beholder is to breathe the aura of those mountains, of that branch. In the light of this description, we may readily grasp the social basis of the aura's present decay.

Walter Benjamin, "The Work of Art in the Age of Its
Technological Reproducibility" (1936)

Combine the endless space of air, the sun's warmth, the clouds that are described to my understanding spirit, the frequent breaking through the soil of a brook or the expanse of the wind-ruffled lake, the tactual undulation of the hills, which I recall when I am far away from them, the towering trees upon trees as I walk by them, the bearings that I try to keep while others tell me the directions of the various points of the scenery, and you will begin to feel surer of my mental landscape. The utmost bound to which my thought will go with clearness is the horizon of my mind. From this horizon I imagine the one which the eye marks.

Helen Keller, *The World I Live In* (1909)

An aura-to-go? Are you sure that is what we should be looking for?

Madeline Gins, *Helen Keller or Arakawa* (1994)

Jayne Lewis, "'The Endless Space of Air': Helen Keller's Auratic Worldbuilding" in: *Imagining Air: Cultural Axiology and the Politics of Invisibility*. University of Exeter Press (2023). © Jayne Lewis. DOI: 10.47788/HHPU4766

Two Horizons

Melancholia pervades one of Walter Benjamin's most cited essays, a well-wrought elegy for the "aura." For Benjamin, "aura" was the penumbra of luminous authenticity that once haloed "the work of art," only to "decay" under the modern regime of "technological reproducibility"—post-Enlightenment routines of "mechanical reproduction," as the more common translation of his title has it. These, in his view, were exemplified by photography, although they applied also to the very print medium within which aura's decay is documented, lamented, reproduced, and disseminated. But Benjamin also apprehends, peripherally, another dimension of aura; if it is primarily a visual phenomenon, the word "aura" itself nonetheless means breeze. Corresponding associations with breath and air identify it with the ambient and the invisible: "a strange tissue of space and time." Inhabited not by art but by humans, does the aura so automatically sink into "decay"? What beyond inexorable "social" forces might it help us "readily grasp"?[1]

These questions go unanswered. Even reconceived as air, Benjamin's aura retains visual qualities of shadow and horizon, markers of distance, and the highly aesthetic atmosphere that his essay creates is downbeat, abstracted, as far away as it is near. We might contrast it with what arises in the deaf-blind American memoirist Helen Keller's representation of her perceptual practice. Keller too registers proximity as distance. Here again, a horizon. But the accident of double disability has subtracted not only the sense of sight that underwrites Benjamin's lament, but also that of hearing, leaving Keller's "I" to convey "my" experience as a "tactual undulation" within a "mental landscape." As Keller's "I" slips through "the endless space of air," reconfiguring it as she goes, the lexical forms that mark "my" movement are neither nostalgic nor melancholy: they are creative and didactic. Still, their lesson's grasp is tentative. Should "you" combine the elements of Keller's motion-made air, "you will then [only] *begin* to feel surer" of her experience. Which is, in any case, disorienting. "Bearings" may or may not be kept; "directions" do not necessarily direct. Air itself? An interrupted surface. Already ruffled by wind, bodies of water "break through." There are people besides "me" here—"you" (me?), the reader; the "others" who "tell me" things—but the "space of air" is finally a place in "my mind." Only such paradoxes, sustained, render air's horizon limitless, allowing Keller's sightless "I" to "imagine what the eye marks."[2]

One benefit of an airspace at once so pressing and so dynamic is that for Keller's reader—you and/or me—her "endless space of air" feels as palpable as it does notional. A "strange tissue" akin to Benjamin's aura arises from notations of bodily location that fill the gap between "my mind" and others' eyes. If Benjamin laments the auratic tissue's tearing, and elsewhere its stiffening into artifice through technologized routine, in Keller's invisible and yet perceptible hands, language's technique is to sustain, augment, and even regenerate embodied spacetime's intricate weave. Words' surface dislocations

ultimately deepen sensation, opening novel [kin]aesthetic possibilities fore-
closed in Benjamin's essay.

Today, that essay is most likely to be read in *Illuminations* (1968), a high-
theory volume whose posthumous title signals the visual protocols that
contribute to the melancholy "aura" of many of its essays, including "Work
of Art" (1935). Keller (1880–1968), Benjamin's far longer-lived contemporary,
brought her "endless space of air" down to earth in *The World I Live In* (1909).
This second and most self-analytic of her twelve popular books—most
memoirs, all formally inventive works of literary performance art—has long
been of interest to phenomenologists from William James to Maurice
Merleau-Ponty and Oliver Sacks, as well as to recent historians of psycho-
aesthetics and psychosocial architectures. *The World I Live In* itself features
not one but three chapters on "Optimism."[3] Ever a recycler of her own texts
and sometimes, scandalously, those of others—Keller would later expand
these chapters into a book of that title, even as her friend and biographer
Van Wyck Brooks glimpsed in her the "auroral openness" James attributed
to true mystics and saints.[4] Keller's optimism was in some ways a literary
imitation of an outlook routinely summoned because it was expected. Still,
to "optimize" is also to expand possibilities. And it's in this sense that Keller's
lexically located, variably repetitious representations of the "space of air"
impart unique lessons about the air and its diminishing prospects in a textu-
alized modernity. In so doing, these same representations tender an alternative
to Benjamin's elegant materialism, and thus to his melancholia.

Perceptibly shaped by disability, Keller's writing "about" the endless space
of air that circulates about *it* demonstrates the surprising relevance of an
expansive and self-reflexive textual *aisthesis* to a developed world's habits of
being "in the air." For Keller, the technologically grounded representations
that bound her practically and imaginatively to others could not, in her
estimation, be separated from fully human experience and identity. While
the ableism implicit in this view has raised hackles in present-day Disability
Studies, Keller's explicitly aeriform "self-recordings" have nonetheless often
supplied inspiration, and even practical information, to modern and postmodern literary, visual, and performance artists.[5] Coextensive with Keller's
own performative and didactic interactions with the air about her, their work
traces an ethical and unapologetically aesthetic science of air. This "aestheth-
ical" science regenerates airspace for an expanded and infinitely diversified
human being even as human beings gasp for breath under technological and
industrial regimes of air conditioning. In the phrase of one of her most
sensitive postmodern mediators, Keller's work rehabilitates the "spacetime
that gets killed or used up by the species as a whole."[6]

Modeling Loss

Nella Braddy's foreword to Keller's 1930 memoir *Midstream: My Later Life*
noted with satisfaction that "she ... has caused a great disturbance in learned

minds, for she has a disconcerting way of upsetting nearly all preconceived theories."[7] So before entering Keller's "endless space of air," it's useful to consider one of the "preconceived theories" of air that it upsets. In brief: Benjamin's melancholy "description" of aura as an imperiled aesthetic phenom-enon is one in a fundamentally nostalgic, if linguistically sensitive cluster of modern and postmodern critical narratives concerning air's fate in modernity. All embrace the death-by-science model most visibly championed by Max Horkheimer and Theodor Adorno in their influential *Dialectic of Enlightenment* (1944).

This model is anticipated in Leo Spitzer's classic essay "Milieu and Ambiance" (1942). Like Benjamin, Horkheimer, and Adorno, Spitzer was a German Jew in exile from Hitler's technocratic regime of terror. These conditions, perhaps, induced him to deplore what his essay measures: Western modernity's distance from an ancient Hellenic conception of "atmosphere." *Aer ambiens* is the "name" that the early Greeks "invented for the ring or orb of vapor ... supposed to be exhaled from the body of a planet and to be part of it, which *air* itself was not considered to be.'" This distinction preserved atmosphere's auratic, intimate, and sheltering qualities while leaving "air" a separate matter for purposes of practical bodily use. "With the progress of science," the word "'air' came to refer only to the supposed limited aeriform environment of the earth or any other planetary or stellar body." While the "'surrounding air' [still] reminds us of *aer ambiens*," that gaseous, quantifiable envelope *is* finally but a memento, one that also marks our distance from *aer ambiens*. That distance makes it impossible to experience our surroundings directly; instead we rely on mediation, as supplied by technology.[8]

Speaking of mediation, meaningful and retentive speech was, by Spitzer's lights, once reciprocally embedded in *aer ambiens* and valued by human culture as such. But it finds itself increasingly depleted and alienated into text—a visual sign system consisting of objective shapes (letters) systematically repeated and acquiring meaning according to the measurable spaces that divide them. Meanwhile, animating figures of speech once intimately involved with aeriform referents—Spitzer's reader might think of "inspiration" or "aspiration"—deracinate. Deflated into inert tropes, they are no longer capable of dynamically placing human beings in their in-breathed or out-breathed environment. Seeking a perpetrator of these changes, Spitzer fingers the scientific and technological Enlightenment of the eighteenth century, which allegedly flattened air into an object of analysis, classification, and domination. Reified and dispersed in print, language's characteristic, sight-based and sound-echoing abstractions, delineations, propositions, referential designs— the "light of ... description" Benjamin summons in *Work of Art*—further mark the loss of *aer ambiens*. But these same linguistic elements also perpetuate that loss as a sign of mastery, and it is this paradox that pervades formal critical discourse. Operating within that discourse's "milieu," Spitzer's own "Milieu and Ambience" is representative of its modern compulsion to answer the idiom of science with the pathos of depletion.

In our own digital era, Hermann Schmitz has influentially repeated Spitzer's downcast history of air while reversing its defining episodes. If Spitzer's primordial Hellenic atmosphere was sequestered from plain "air," Schmitz's synthesized the inner and outer "worlds" of ancient air's human inhabitants, who were also its constituents; the (idealized) ancient body's sense of the surrounding world coincided with its sense of itself as a "perceptible *felt body*" in that world. "Emotions" and "impressions" arose as "atmospheres," integrating "objective" surroundings with the human body that sensed their textures, variations, and pressures. Euclidean geometry "split" this one "world," cleaving it into inner and outer "spaces." Subsequently, mastery of each has depended on its oppositional distance from the other. "What remain are inner worlds for self-control and an external world for mastery of the world, first by God, then by humans and their works."[9]

Schmitz decides that classical geometry is responsible for this rupture insofar as it instilled a "locational" scheme in which "areas exist in space."[10] They do so in such a way that "areas, lines, and sections"—all modes of separation and sequestration with close analogues to alphanumeric writing systems—become essential to the perception of three-dimensional bodies as such. This conceptual framework valorizes distance and excludes the possibility of a body that projects its symbolic world by proceeding through it. Schmitz nonetheless hunts for "spaces without area" that might still exist as what Peter Sloterdijk, charting a similar downward spiral accelerated if not precipitated by Enlightenment science, conceives as malleable spheres.[11] "Spanned by ... movement," not measurement, Schmitz's own atmospheres manage to survive the tyrannies of modern technologies of mediation as "spaces without area" which are "experienced through rhythmic and tonal movement."[12] Here the "*felt body* of a person" is "the epitome of everything that he himself can perceive as belonging to himself, in the vicinity—not always within the boundaries—of his *material body* [*Körper*], without making use of the five senses of sight, hearing, touch, smell, taste." Its sensory impairments sublimated into choices, this atmospheric being "goes beyond the boundaries of one's own felt body and connects felt bodies to one another, as well as to *incorporeal* figures."[13] But if it rescues some element of an archaic super-sensory atmospheric body, Schmitz's story also distances it from the verbal, conceptual, and scientific technologies of enlightenment, bound as they are to the discrete representational "figures" typical of both geometry and writing. "The" body, ironically enough, disappears as well. For all their promise, Schmitz's "atmospheric spaces" thus sustain the dispiriting divisions that they seemed to eliminate.

Remodeling Loss

"Your condition allows you to escape ... 'the ennui one feels regarding this too solid and heavy world.'"[14] Quoting a Hamlet-quoting Stéphane Mallarmé, the postmodern poet Madeline Gins's second-person meditation on, in, and through Helen Keller's "I" exempts its lightweight "you" from melancholia's historically prestigious oppression.[15] Is this merely an idealization of Keller's unchosen "condition" of deaf-blindness? Surely cause for melancholia: At the speech- and thus humanity-acquiring age of nineteen months, Keller permanently lost the two of Schmitz's "five senses" most closely implicated in symbolic language. Furthermore, Keller became visible to herself as a human subject of communication—conventionally marked as "I"—through a cumbersome and numbing suite of mechanized post-Enlightenment writing technologies that could lock her writing into a citational spiral: "Finger spelling" that has since been likened to the programming of artificial intelligence; wooden frames that disciplined her child hand to reproduce a mechanical font-like script; bespoke typewriters; the fixed stylizations of the recently engineered Braille compositor.[16]

These same mechanical and mass-produced devices, however, also made communication a matter of mobility and tactile sensation in three-dimensional space, and it was through them that Keller produced a vibrant and realistically hopeful archive of so-called self-recordings. Incubated in the technology that Benjamin conscripted in order to conceive and communicate the cause of the aura's decay, photographs of Keller often showed her facing forward toward such instruments. An exceptionally photogenic photographic subject, Keller lends "auroral openness" to both the machines and the photographs, often seeming to transmit this quality from alternative "auratic" media: old books, visual images, sunlight itself (see Fig. 9.1). Such photographs in turn cross-reference similarly [com]posed photographs (see Fig. 9.2 and Fig. 9.3), in which Keller reaches toward natural objects (flowers, dogs, horses), other people (Dwight Eisenhower, Winston Churchill), and works of art (medallions, friezes, sculptures).[17] Keller's fingers only lightly skim the surfaces of such objects, however. As in the flight of unsighted fingers over lines of Braille—a reading practice evocatively documented by Marta Werner as a component of Keller's poesis of invisibility—the "space of air" seems to open just at what we would expect to be a point of fixed contact.[18] This renders both the photographs and the self-recordings often printed alongside them dynamic diagrams of air. They render air visualizable if not visible.

Vilém Flusser traces photography's genetic line to writing's "graph"; both are "technical images," "image[s] of concepts" that transmit meaning across surfaces simultaneously abstracted from and projected back into space and time.[19] Keller's lexical self-recordings, almost always supplemented with some of the photographs in question, routinely reproduced the word "air" in a similarly trans[re]ferential fashion. This holds truest in *The World I Live In*. Keller famously chafes against public obsession with her unusual sensory "world" to the exclusion of "the" usual, consensual world itself:

> While other self-recording creatures are permitted at least to seem to change the subject, apparently nobody cares what I think of the tariff, the conservation of our natural resources. ... My editorial

Fig. 9.1. Helen Keller seated alone and typing at Radcliffe College (1900). Reproduced courtesy of American Foundation for the Blind (Helen Keller Archives).

friends ask for an account of grown up sensations. … Until they give me the opportunity to write about matters that are not me, the world must go on uninstructed and unreformed, and I can only do my best with the one small subject upon which I am allowed to discourse.[20]

The product of a double negative barred from changing "*the* subject," Keller's "small subject" is a sensory being every bit as much as it is a lexical one. In the always already formally mediated "world" this subject builds, "sensations," perceptible between letters and lines of writing, are not *only* its. They are

Fig. 9.2. Helen Keller smelling and touching cut flowers (1919). Reproduced courtesy of American Foundation for the Blind (Helen Keller Archive).

both subjective and objective, communicative and receptive, peculiar and extended into multiplicities of others.[21] Keller's "I" and its sensations dwell in the air that carries all of these capacities.

Of its relationship to its surroundings, Keller's "one small subject" has this to say: "Every atom of my body is a vibroscope."[22] Sensitive readers have been drawn to this metaphor as a figure for Keller's intersubjective, "tactual" way of knowing and communicating her otherwise incommunicable world. But in the context of the technical imagery that includes literature, the vibroscope

Copyright, 1907, by The Whitman Studio

The Medallion

The bas-relief on the wall is a portrait of the Queen Dowager
of Spain which Her Majesty had made for Miss Keller

Fig. 9.3. The Medallion. *The World I Live In*.

surpasses its own metaphor. A now forgotten but literally eloquent forerunner of today's seismographs and electrocardiographs, the vibroscope was a writing technology that gave air visible form by transferring vibrations onto a paper surface. In its time, the vibroscope thus gave scientific literature a way to register its own designs, especially when that writing was also didactic. For example, an 1896 physics textbook presents the vibroscope as a "graphic method of studying sounds."[23] A stylus, mounted on a rotating cylinder, responds to vibrations—agitations of the surrounding air. The air transfers the resulting gyrations to "smoked paper," expressing them there as shapes and lines, a kind of nonrepresentational, autogenic, impersonally first-person writing. "Smoked paper" reveals air as supervalent form—form that emits its own content as form as it moves. The paper's expressive surface doesn't significantly differ from the air around it. In turn, in Keller's hands, the figure *of* the vibroscope, continuous with the instrument "itself," helps "you" imagine "air" not just as a space or chemical entity but as an equally internal and ambient phenomenon, one whose "accumulations" and "concussions" spontaneously arise from a play of spontaneously delineated surfaces. Such effects redouble as a new form of multiplicitous human being when vibroscopic sussings become an "I."

The vibroscopic "I" that both builds and sounds *The World I Live In* is an objectified, fully textual avatar of the protagonist of Keller's earlier, bestselling memoir *The Story of My Life* (1903). This first-person "story" often shades into the third person, blurring its difference from contextualizing letters and commentary from Keller's closest friends. In *Story*'s defining plotline, loss of sight and hearing precipitated four years of existence as what Keller's final memoir, *Teacher* (1962), would call "a Phantom living in a world that was no-world." This almost purely atmospheric being had "no sense of time," only of surrounding space in which it could not distinguish between its own body and those of others, human and inhuman.[24] While Keller did not idealize, romanticize, or sentimentalize Phantom's atmospheric body, she did canonize its transition into the social sphere of human language when Keller's legendary teacher Anne Sullivan traced the letters W A T E R into one of Phantom's palms while pumping water into a mug grasped in the opposite hand.

Besides producing a reproducible link between signs, objects, and ultimately concepts, the water "miracle" opened an atmospheric stream of associations, culminating in the production of a linguistically fluent personal identity: "As the cool, fresh stream burst forth, Teacher made me put my mug under the spout and spelled w-a-t-e-r," Keller (age twelve) wrote in a precocious magazine sketch. "Water! The world startled my soul, and it awoke, full of the spirit of the morning, full of joyous, exultant song. Until that day my mind had been like a darkened chamber, waiting for word to enter and light the lamp, which is thought."[25] *The Story of My Life* expanded, multiplied, and dispersed this point of transition:

> Someone was drawing water and my teacher placed my hand under
> the spout. As the cool stream gushed over one hand she spelled into

the other the word water, first slowly, then rapidly. I stood still, my whole attention fixed upon the motions of her fingers. Suddenly I felt a misty consciousness as of something forgotten—a thrill of returning thought ... I knew then that "w-a-t-e-r" meant the wonderful cool something that was flowing over my hand. That living word awakened my soul, gave it light, hope, joy, set it free![26]

In this endlessly recycled anecdote, the manual alphabet turns a concept—water—into a percept. When the anecdote is reprinted, a mass-produced mug that initially held w-a-t-e-r becomes the organic container of the "hand." Inseparable from the conceptual and tactical navigation of letters, an atmospheric effect of sensed abstraction is deepened through a kind of typographical "mugging"—mechanical duplication in representational space.

The water miracle became a signature in cinematic remediations of Keller's "story," most notably William Gibson's *The Miracle Worker* (1962). But it also epitomizes the manner in which Keller's doubly fluent (verbally eloquent *and* free-flowing) "I" transmits air's textures, temperatures, tones, vibrations, smells, curvatures, and degrees of moisture and pressure throughout its literary self-recordings. Reversing a tragic history of air that splits concept from experience, interior from community, human beings from their ambiance, "air's" definable and even abstract qualities are inseparable from their qualitative effect on the freeform body that perceives them.

Repeatedly articulated in subtly transforming print, Keller's air builds three vital continuities. For one thing, air mediates between Keller's mental world and her sensory one. In *The Story of My Life*, Keller's "I" thus speaks of the "atmosphere of jostling, tumbling ideas I live in."[27] Second, air circulates between Keller's "I" and her sighted, hearing readers, a common currency. Consequently (and coincidentally), Keller's implied readers can experience the areas adjacent to their bodies as adjacent also to hers, an effect enhanced when close friends served as first readers and active reproducers of Keller's story and image.

Third, metaphor and denotative language, like fantasy and fact, breathe the same air. Keller's self-recordings use "air" to realize her "I"'s experience in space and time for her reader: "Air" can be "low-hanging" or feather light, "pungent" or sweet, depleting or enlivening depending on where (Manhattan, say, or Keller's native Alabama) and when (morning versus afternoon) it is; summer comes across when "air" takes the shape of "heat waves." But dreams are also "castles in the air." Likewise fairy tales: Keller's first published piece was the atmospherically entitled *The Frost King*. The fairy tale, notoriously, faced successful charges of plagiarism, and Keller's authenticity is questioned today in the ableist Twitter thread #helenkellerfake.[28] But then again, "airs" are also affectations and Keller often cheerfully admitted to "put[ting] on a deliberate air." As the figural and the literal meet in midair throughout her writing, Keller anticipates and indeed models Gernot Böhme's more recent formulation of "atmosphere" as "what mediates objective aspects of the environment with aesthetic feelings of a human being."[29] In *Midstream*, the "tremulous, far-away murmur" of Jascha Heifetz's violin "touched my face, my hair, like kisses remembered and love-lit smiles" while remaining "immaterial, transient as the sigh of evening winds."[30] Decades earlier, Keller's "I" nearly drowned in the waters off Cape Cod, only to find her own gasping body gathered in an "all-enveloping element—life, air, love."[31]

There's no place air is not in Keller's writing, nothing it cannot envelop or become. Whenever "air" materializes as "atmosphere," it thus expresses one of its many potentialities within an immediate rhetorical setting, whose historical determination it both reveals and transcends. For example, Keller's semi-autobiographical 1927 study of the mystic Emanuel Swedenborg pronounces his eighteenth century "the coldest, most depressing time in human history," a period engulfed in a "sinister, oppressive atmosphere."[32] In melancholy modernist air theory, this *is* the Enlightenment that would kill aura and ambience. But for Keller, who absorbed Swedenborg's words as those of a mystical pragmatist (and ironist), his later-life vision of supermaterial human life opened a different potentiality. Embracing it, Keller also embraces Swedenborg's many innovations and practical speculations as a scientist as well as his religious commitment to what "can be published in print and read and understood by whole populations." Swedenborg himself regarded print "technology" as a liberating and uniquely earthly innovation, continuous with an unfolding "revelation ... through writing" and the only "way the divine can flow in from heaven."[33]

A good thought for a compulsively literary mystic in training! Keller's own "air" draws on the language of scientific enlightenment as well as on the scientific Enlightenment's diagrammatic and didactic literary conventions: "I go through continuous space and feel the air at every point, every instant. I have been told about the distances from our earth to the sun, to the other planets, and to the fixed stars. I multiply a thousand times the utmost height and width that my touch compasses, and thus I gain a deep sense of the sky's immensity."[34]

World

Keller's aeriform "I" revises a phantom into a proleptic figure of expanded, diversified, ecologically involved human being infinitely reproducible in the imaginative perceptions of others. Keller's "I," like Phantom, relies on two senses: the "tactual," which encompasses touch, proprioception, vibrational perception and balance, and the olfactory. Inscribing these senses concurrently within repetitive writing practice and within the social and natural world through which her physical body moved, Keller's often surreal self-recordings reconstruct both as equally receptive *and* communicative, indeed representational. This moves them to the same plane as the outwardly lost senses of sight and hearing, permitting what Keller termed "analogies of sense perception" to move freely across all four of these senses throughout her writing.

A blind subject who can smell might avail itself of visual metaphor. "I can imagine what light is to the eye," Keller wrote. "It is not a convention of language, but a forcible feeling of reality, that at times makes me start when I say, 'Oh, I see my mistake!' or 'How dark and cheerless is his life!' I know these are metaphors. Still I must prove with them."[35]

Keller's "I" "prove[s]" a socially interactive internal atmosphere, one replicated in the alphabetical particles, gathering into words and printed lines, that constituted her writing and made her inner life shareable. In a lexically registered mind-space that is also a trace-diagram of the senses, "all creatures, all objects pass, and occupy the same extent there that they do in material space."[36] Keller's "I" can thus "declare that for me branched thoughts, instead of pines, wave, sway, rustle..." Such dislocations prove viable and even reproducible alternatives to melancholia and the historical paradigm of irreversible loss that motivates it, but that melancholia also sustains.

Nowhere are Keller's internal atmospheres more perceptible, or more explicitly articulated, than in *The World I Live In*. This self-recording originated not in biological life but in an earlier text, a group of impressionistic and discontinuous pieces that Keller first published under the title *Sense and Sensibility* in *The Century*, one of the most widely circulated and intermedial literary magazines of the period. While evocative of both body (sense) and body world (sensibility), Keller's (not) original title also cites Jane Austen's 1811 novel of the same title, a shadowed comedy of manners that ironically distances itself from Enlightenment fictions of sensation. Within months, Keller had replicated and expanded—amplified—her own *Sense and Sensibility* pieces into an impressionistic assemblage of playful phenomenological reflections that revives the ludic, fragmentary textual synesthesias of such haunting and anachronistic Swedenborg-era pastiches as Laurence Sterne's *Sentimental Journey* (1768) or Henry Mackenzie's *Man of Feeling* (1771). Through this literary and yet kinesthetic method, *The World I Live In* seeks to teach its reader how to move through its narrator's sense world.

If procedurally generated, this "world' is also partly composed of precirculated tropes and quotations, some Keller's own, that were already "in the air," as it were. Meanwhile, the senses make up for one another's work to exude the trans-sensory, pan-sensory, vibroscopic "I" who speaks the book—literally the "world I live in"—as a whole. Four chapters feature the "seeing hand" and its world-building "sense of touch." One takes up "vibrations" as a whole, one smell, three dream life. "A Chant of Darkness" in verse is adapted from the Book of Job, and Keller's previously published "story" of her transition from Phantom to Helen is reprinted after philosophical reflections on optimism both "without" and "within."

Its parts circumscribed within the common medium of language, this composite structure makes concepts sensible, sensations conceivable, and both transmissible. "You, dependent on your sight, do not realize how many things are tangible," Keller's "I" reminds me. Among such "things" are abstractions, such that what touch actually perceives need not be, say, texture or temperature. It can instead be "beauty … derived from the flow of curved and straight lines which is over all things." Keller's "I"'s vibroscopic hand makes beauty tangible by making it communicable; "lines" become sensation. Throughout, "the silent worker is imagination," which uses "associations, sensations, theories" to realize the "spaces" Keller's "I" moves through as co-created environments. The inevitable mediation of sensation produces a reality that lays claim to objectivity—hence Keller's perceived reality gets to count as a "world." "'What does the word line mean to you?' you will ask," predicts Keller's "I." "It means several things." Yet the working imagination finally strikes. "The word line's" full range of meanings remains available only to Keller's "I": "every object appears to my fingers standing solidly right side up, and is not an inverted image on the retina, which, I understand, your brain is at infinite though unconscious labor to set back on its feet."[37]

Keller navigated material space awkwardly. Braddy reported that "her sensory equipment is in no way, except perhaps in the sense of smell, superior to that of the normal person. She seems totally without [a] sense of direction, … frequently starts toward the opposite wall instead of the door, and orients herself by contact with the furniture. When the rugs are taken up she is completely bewildered and has to learn the whole pattern again. Her sense of distance is also poor."[38] The operating instructions that "I" provides "you" call for constant dislocation and perpetual, ever but temporary reorientation. Carried forward simultaneously in a perpetually displaced and self-decomposing linguistic field, the resulting psychomotor experience builds a world. Pushed into ever smaller force fields, Keller's "one small subject" discovers free space within technologized routine. Hence Keller's reader (my I, her you) is thrown off balance, with her "so accustomed to light, I fear you will stumble when I try to guide you through the land of darkness and silence."[39]

In life, Keller guided performance artists such as Charlie Chaplin and Martha Graham, both of whom adapted line-breaking, convention-disrupting movement techniques from her. Filmmakers initiated their own incoherent dialogues with her. A surreal 1919 silent film scripted by the Civil War photographer Francis Trevelyan Miller, *Deliverance*, lurches between a jerky reenactment of Keller's early years, a scene in which "she" teeters on the edge of a cliff outside a cave where allegorical figures on a high cliff wrangle for her soul, and footage of the "real" Keller swerving through the air in a biplane.

Keller airily pronounced *Deliverance* "hilarious"—and wryly complained only that "My Wrentham house, which is historic, does not appear in the picture."[40] But a different house does appear in Nancy Hamilton's Academy Award-winning 1954 "picture," *Unconquered: Helen Keller in Her Story*. Heavily indebted to gothic film techniques recently enabled by technological advances in sound and lighting, this "documentary" places Keller at once "in" her two-storied house and "in" the film story. Both are rigged and both are phantasmic. Disembodied voice-over narration by Hamilton's lover Katherine Cornell haunts a nervous camera that tracks Keller daily movements inside

and outside Arcan Ridge, the Connecticut house custom-designed for her
in the 1940s. Rigged with rope lines that make Keller seem to move like an
ungainly trapeze artist up and down stairs and walkways, the house is also
a world of ambiguously sourced shadows and voices no one hears. (Unanswered
letters on a desk whisper among themselves; the deaf Keller's companion,
Polly Thomson, confesses that she speaks aloud to "drive away the silence.")
Capturing the "atmospheric body" of the ever-still-present phantom,
Hamilton's "Helen" wavers on a balcony beside a Braille thermometer. "The
first concern of the day is always the weather," Cornell's voice brightly reports.
"Helen" lifts a finger to the air. "Clear and cold!" the voice-over carols, from
everywhere and nowhere.

Diana Fuss observes that after a gas furnace exploded, Arcan Ridge,
atmospherically enough, went up in flames; though the original house was
designed to channel vibrations into Keller's body, its all-electric, "fully auto-
mated" reconstruction "radically undermined [Keller's] tactile and olfactory
knowledge."[41] Though Fuss rightly names Hamilton's "reverential" determination to "enter directly" into Keller's "intimate"—vibrational—life, the film
medium's "endless" loop of projected imagery may be more aligned with the
air about her. Keller's image teeters through that air as she makes her way
up stairs or maneuvers herself along an outdoor leadline.[42] Yet in their tenta-
tiveness, all of these movements produce new, polycentric modes of being
within canonical household and generic space—psychomotor techniques that
resist deterministic repetition even as they extended life into the film medium.

This potential eventually attracted the Japanese conceptual artist and
procedural architect Shūsaku Arakawa (1936–2010). Arakawa's lifelong
dialogue with Keller culminated in the "reversible destinies" project he under-
took with his wife Madeline Gins in the early 2000s. Gins's verbal recordings
of that dialogue in her 1994 prose poem *Helen Keller or Arakawa* build an
aptly literary rendering of a project whose ongoing premise was that perception is conditioned and indeed determined by technological, social, and
conceptual regimes of repetition. The dislocating physical and representational
spaces that thwart habit also undo these regimes. As they "construct sensoria
that will elude mortality," they potentially sustain human life at once within
and beyond their own charted limits.[43]

Arakawa playfully and hopefully explored this possibility across multiple
media whose often cryptic humor is always sanguine rather than melancholic,
as practical and immersive as it is abstracted and theoretical. With Gins, he
collaborated in the design of several still-extant apartment buildings that, in
the Arcan Ridge mode, exploit undulant yet static wave-like surfaces, vibrant
color, and the architecturally atypical geometric forms of spheres and tubes
to continually challenge and stimulate the senses, forcing individual body-
minds to free themselves from their limitations and thereby potentially
reversing destinies inscribed in the very air around them. Arakawa and Gins's
playful Tokyo apartment complex "Reversible Destiny Lofts Mitaka: In
Memory of Helen Keller" is a still-living move against oft-rehearsed

atmospheric attitudes of mourning and melancholia. Implicitly remembering Keller in its uneven floors and off-kilter appliances, the complex even honors the role that "memory"—functionally indistinguishable from forgetting—literally plays "in" the experience of present space.

Gins's 1994 prose poem *Helen Keller or Arakawa* returns these dynamics to literary form. It is a dense meditation on several of Keller's self-recordings but especially *The World I Live In*. Marginal citations of relevant conceptual works by Arakawa such as his mylar screen print Alphabet Skin (1966) deliberately withhold images of these works, suspending the kinetic diagrams that Gins's text plots in language whose intricate puns meld psychomotor movement with the formal geometry of printed characters. The "I" that speaks in Gins's speculative fiction is a composite of Gins, Arakawa, Keller, and several more obscure didactic figures, including one called Plenum, that only possibly correspond to organic bodies. The impossibility of determining where one stops and the other[s] begin reiterates the impossibility of determining whether Keller's deprivation of vision and hearing is the blank background of Arakawa's diagrammatic art or vice versa. The resulting lexical experiments teach the perceptual mind to generate novel "positional permutations"; these in turn potentially relay the conceptual ground for the erasure of distinction between person and "surround." An architectural body is therefore an "organism that persons."[44]

These procedures explicitly remodel air (predetermined, objective) as atmosphere (fluctuant, amenable, and procedurally generated). "One aerial ocean smacks into another and a sense of appearance arises"; an "air and aerated world [is] opened up by names and contours," then "in conjunction with arrows and numbers."[45] "You" are enticed to "stare specifically into 'thin air,' that medium, and record those lapses of attention which constitute blanks" or to contemplate an invisible leaf whose rib and tendril "se[t] up the pattern of enactment for all subsequent drawing-throughs" back into thin air.[46] Meanwhile, an "I" that seems likely to be "Arakawa" speculates that "'It may be moves to make adjustments of scale, demonstrably part of the thinking process, require full atmospheric mindbody articulation. This is what I believe."[47] To entertain this playful credo might be to open, generate, or perhaps discover, new space within Keller's own writing. Positioning the text of *Helen Keller or Arakawa* in the interstices of Keller's aeriform writing, Gins/Keller/Arakawa holds that "I had to learn to open up that significant yet barely perceptible bit of spacetime between mug and the liquid it contained."[48] What—or where, or how—could this "bit" be but Keller's own endless space of air?

Fallen Angel

Keller's idiosyncratic "tactual" style both models and conducts a relationship to the air that makes destiny itself seem reversible. Braddy noted that Keller's "sensory equipment is in no way, except perhaps in the sense of smell, superior to that of the normal person."[49] In a pivotal chapter of *World* entitled "Smell, the Fallen Angel," that "equipment" takes on a life of its own.

Uniquely positioned between common noun and transitive verb, "smell" is most affined with air; its chemical structure, like air's, is molecular rather than vibrational.[50] Consequently, smell can materialize the abstraction of passing time: "Odors, instantaneous and fleeting, cause my heart to dilate joyously or contract with remembered grief."[51] Yet this cardiopulmonary response is not to grief itself but to grief *remembered*; certain odors don't revive the past but rather make it present it as absence. At home in what Keller rather chillingly calls the "all-encasing air," the sense of smell is thus *World*'s most typical expression of Keller's vibroscopic capacity for "varied, instructive contact with all the world, with life, with the atmosphere whose radiant activity enfolds us all."[52] Points of olfactory contact potentially deflect the stories of alienation, isolation, and loss charted on the theoretical and communal levels by Benjamin, Spitzer, Schmitz, and Böhme. These stories are diverted into atmospheric—equally shared, individuated, sensed, and notional—experiences of "concussion" and "accumulation."

Though her chapter title is a personification, Keller's olfactory "I" dwells on "person-scent," "person-odor," and "person-smell" as these both convey and are the "impression of [a] person."[53] Keller's human aromas coincide with the first industrial syntheses of perfume—the mechanical production of "fake" aura—but as literary figures they also converge with her contemporary, the sociologist Georg Simmel's 1907 essay "Sociology of the Senses." Simmel draws on aura to both figure and denote the ways in which individuals both represent and recognize themselves in the "atmosphere" of social sensibility. Spheres of air that arise from individual bodies render humans' perceptions of one another qualitative, aesthetic, and intersubjective to the extent they are also ultimately impersonal: A "surrounding layer of air scents every person in a characteristic way." Hence, "that we smell the atmosphere (*Atmosphäre*) of something is a most intimate perception of that person, that person penetrates, so to speak, in the form of air, into our most inner senses."[54] In a later essay, Simmel anticipates Benjamin: "the perception of people is … like the perception of landscape." In both perceptual experiences, "atmosphere" is "at once the milieu into which we enter and in which we are contained and the … intermediate third which connects and unifies the two sides of perception."[55]

"Smell the Fallen Angel" tests the two-sided (and interpenetrative) human atmosphere as a practical maxim when its "I" literally moves among the "exhalations" of various persons. But the chapter's very title is also a personification, a figural device that removes smell from the physical atmosphere and invites us, factitiously, to visualize smell as person. Any "fallen angel" will conjure an iconic image from the early stages of the technological Enlightenment: Albrecht Dürer's obsessively geometric Melencolia I. In the nostalgic mode, Heather Keenleyside finds that Enlightenment literature degrades personification, once the most animistic and vital of rhetorical tropes, to a dead form: "Cut off from myth and authentic animism," it "survived only as a conventional device."[56]

It is from this side of Enlightenment that Keller's "fallen angel" opens new possibilities and environments for experiencing the functional, fictional bodies we call personifications as if they were persons. What looks like a personification is not quite; after all, an angel is not a person, dead or alive: It's an aeriform messenger, one that takes human form only when it falls into representational—geometrical—space, such as the one that Dürer fills with technical instruments. Keller's fallen angel, as invisible as blind Milton's literary, eloquently air-manipulating Satan, cannot fill space, only communicate [through] it. Just so, smell isn't actually likened to a fallen angel; rather, "there is something of the fallen angel *about it*."[57] A seemingly exacting preposition solicits us to experience the figure of the fallen angel through what hovers around (about) it as Keller's "I" communicates something "about" it.

Though Keller drolly observes that "for some inexplicable reason the sense of smell does not hold the high position it deserves among its sisters," it remains "in my experience ... most important."[58] Smell breaks down to a kind of pneumatic mnemonic, turning event into experience. "I never smell daisies without living over again the ecstatic mornings that my teacher and I spent wandering in the fields, while I learned the new words and names of things."[59] Daisies are, of course, virtually scentless. But especially when represented as letters, their non-scent expresses smell's freedom from the laws of common sense. In much the same spirit, Keller brings forward scent's capacity to disorient and delocalize within present space: "The finest whiff from a meadow ... displaces the here and now. I am back again in the old red barn ... Again I touch Brownie."[60] Keller's "whiff" is somatized yet superfine: The word itself combines the action of quick inhalation with what "is" inhaled.

In the natural world, smell never replaces sight and hearing. Rather it shares their horizon. Thus, "earth odors" that temporarily rise after a storm meet both three-point visual perspective and musical chord. In their shared evanescence, all have the same effect: "As the tempest departs, receding farther and farther, the odors fade, become fainter and fainter, and die away beyond the bar of space."[61] This "bar of space" sounds at first like a limit. But it could as easily be a musical phrase or a photographic phase. As such, that "bar" is part of the "endless space of air" but not the contradictory end of it.

Similarly, death stalks "the world I live in," but is also slightly dislocated from the body that undergoes it: "The other day I went to walk toward a familiar wood. Suddenly a disturbing odor made me pause in dismay. Then followed a peculiar, measured jar, followed by dull heavy thunder. I understood the odor and the jar only too well. The trees were being cut down."[62] Keller's "I" anticipates the asphyxiating deforestation of our own time without waning into a global consciousness of loss, or waxing prophetic. Instead, the terms of this experience remain both intimate and ambient, even when "an unfamiliar rush of air ... told me that my tree friends were gone. The place was

empty, like a deserted dwelling."[63] Pure medium, Keller's "I" floats on the
line between odor and dwelling, composing both as it goes: "I stretched out
my hand. Where once stood the steadfast pines, great, beautiful, sweet, my
hand touched raw moist stumps."[64] The grammatical shift from now-absent
pines to the notionally present hand allows this sentence to say more than
that my hand touched the stumps of now vanished pines, more even than
that there is a correspondence between hand and stump. With "where," the
phrase also realizes the "endless space of air" that both contains and conveys
the pines now vanished, the present stump, and the imaginary hand that
moves between them. Within air's common "space," each is a phantasmic
expression of the others. And yet this consolation:

> An unreasoning resentment flashed through me at this ruthless
> destruction of the beauty that I love. But [… t]he air is equally
> charged with the odors of life and of destruction, for death equally
> with growth forever ministers to all conquering life. The sun shines
> as ever and the winds riot through the newly opened spaces. I know
> that a new forest will spring where the old one stood as beautiful,
> as beneficent.[65]

As fantasies of conquest, such consolations can seem naïve, rote, apolitical.
But if we are seeking a revitalized idiom of atmosphere in this dire
Anthropocene, we might still try to follow Keller's didactic navigation of a
"space of air" that holds both the living and the dead, the accomplished and
the yet-to-be.

"Odors deviate and are fugitive," Keller writes, "changing in their shades,
degrees, and location." As they do so, Keller proleptically transforms
Benjamin's melancholy horizon into something that can still be sensed: "There
is something else in odor which gives me a sense of distance. I should call
it horizon—the line where odor and fancy meet at the farthest limit of
scent."[66] Sympathetic translation exists at and because of this horizon, scenting
new forms of community with new forms of intimacy that include imper-
sonality and phantasmic non-persons. "Odor" itself "seems to reside not in
the object smelt but in the organ," and "since I smell a tree at a distance, it
is comprehensible to me that a person sees it without touching it. I am not
puzzled over the fact that he receives it also as an image on his retina …
since my smell perceives the tree as a thin sphere with no fullness or content."[67]
Keller's "thin sphere with no fullness or content" projects Schmitz's idealized
arealess space from within a geometry that acknowledges loss and admits
phantoms.

"Listening" to the Trees

"Smell the Fallen Angel" features a photograph transferred from the *Sense
and Sensibility* series. Its caption seems to cast Keller as "'Listening' to the

Copyright, 1907, by The Whitman Studio

"Listening" to the Trees

Fig. 9.4. "Listening" to the Trees. *Century Magazine* 1909.

Trees" (see Fig. 9.4). But the caption does not specify Keller as the subject of "'listening,'" as the scare quotes seem to reinforce, and in the image itself, Keller's image literally hovers; not only do we not see her feet, but they also seem not to be touching the ground. It is unclear how far from the ground she is and thus a question arises as to whether this levitating, subjectless phantom is here or not. Is anything? "The Trees" themselves, despite the definite article, occupy no specified location; they dissolve at sight into leaves. There is no perspective. Yet at the level of the leaves—which is exactly where the photograph urges us to be, while also holding us apart from them—we are also in a space whose generative potential waits to be perceived.

In *The Life of Plants*, Emanuele Coccia exercises a "metaphysics of mixture" to picture leaves "suspended in the air effortlessly." Their capacity to create their own environment allows leaves to sustain a "relation between the container and the contained [that] is constantly reversible; what is place becomes content, what is content becomes place. The medium is the subject and the subject becomes the medium."[68] Thus to (tactually?) "grasp the mystery of plants means to understand leaves—from all points of view, and not just from the isolated perspective of genetics and evolution. In them is unveiled the secret of what we call 'the climate.'" Such leaves ripple in (as image) and under (as page, the book's leaf) the textual world Keller lived in. But *The World I Live In* also carries a suggestion not yet heeded by Coccia: The vestigial, elusive human that Keller positions among these leaves cautions against unveiling the secret of what we call the climate.

Coccia finds that "our origin is not in us ... but outside, in open air"; hence "the origin of our world is in leaves: fragile, vulnerable, yet capable of returning."[69] Keller, for all her deciduous optimism, also insists on the persistence of the phantom, the ghost in the machine and the machine in the ghost. "Smell the Fallen Angel" thus subsides on an ecstatic, even mystical note. Without odor, Keller writes:

> The sensuous reality which interthreads and supports all the gropings of my imagination would be shattered. The solid earth would melt from under my feet and disperse itself in space. The objects dear to my hands would become formless, dead things, and I should walk among them, as among invisible ghosts.[70]

"Odor" secures reality, but only tentatively, as a sensibility rather than as an object of sense. Living, and loved, objects drift into "formless, dead things," and thence into "visible ghosts." But this is also the fate of the body that perceives them as such. Indeed, in Keller's *Sense and Sensibility* template, the ghost wasn't any of the "things' around her. It was Keller herself: "The objects dear to my hands would become formless, dead things," she originally wrote, "and *I* should walk among them, as *an* invisible ghost."[71]

This ghost is the condition *The World I Live In* requires in order to be visible. If Coccia's leaves compose the "atmosphere of the world," so does Keller's phantom "I."[72] Afloat in its exquisite alphabet skin, her world's atmosphere includes what it can no longer have or hold, and possibly never did or could. To embrace its reality is to learn to breathe in the endless space of air.

Notes

1   "The Work of Art in the Age of its Technological Reproducibility: Second Version," in *The Work of Art in the Age of its Technological Reproducibility, and Other Writings on Media*, ed. Michael W. Jennings, Brigid Doherty, and Thomas Y. Levin, trans. Edmund Jephcott and Harry Zohn (Cambridge, MA: Harvard University Press, 2008), 23.

2   Helen Keller, *The World I Live In* (New York: Century, 1908), 102. Future citations to the original edition of this book appear hereafter as *World*. For a full and sensitive contextualization of *The World I Live In*, see Roger Shattuck, Introduction to Keller, *The World I Live In*, ed. Roger Shattuck (New York: New York Review of Books Press, 2003), vii–xxxiii.

3   On James's interest in Keller, see Joseph P. Lash, *Helen and Teacher* (New York: Delacorte, 1974), 345–47. See also Maurice Merleau-Ponty, *Phenomenology of Perception*, 1945), trans. Colin Smith (London: Routledge, 2002), 267; Erica Fretwell, *Sensory Experiments: Psychophysics, Race, and the Aesthetics of Feeling* (Durham, NC: Duke University Press, 2020); Diana Fuss, *The Sense of an Interior: Four Writers and the Rooms that Shaped Them* (New York: Routledge, 2004); and Oliver Sacks, *Seeing Voices: A Journey into the World of the Deaf* (Los Angeles: University of California Press, 1989).

4   Van Wyck Brooks, *Helen Keller: Sketch for a Portrait* (New York: Dutton, 1956), 164.

5   On Keller's failings as a "disability superstar" thanks to her embrace of ableist ideologies, see Kim E. Nielsen, *The Radical Lives of Helen Keller* (New York, New York University Press, 2004), 8–122; and for a nuanced first- and second-person exploration of those failings, Georgina Kleege, *Blind Rage: Letters to Helen Keller* (Washington, DC: Gaulaudet, 2002).

6   Madeline Gins, *Helen Keller, or Arawaka* (New York: Burning Books 1994), 234.

7   Nella Braddy, Foreword to Helen Keller, *Midstream. My Later Life* (Garden City, NY: Doubleday, 1930), xvi.

8   Leo Spitzer, "Milieu and Ambiance: An Essay in Historical Semantics," *Philosophy and Phenomenological Research* 3 (1942): 188. https://doi.org/10.2307/2102775.

9   Hermann Schmitz, "Atmospheric Spaces/Espaces atmosphériques" (2012), trans. Margret Vince, *Ambiances* 2016: 2–3. https://doi.org/10.4000/ambiances.711

10   Schmitz, "Atmospheric Spaces," 2.

11   Peter Sloterdijk, *Bubbles: Spheres Volume I: Microspherology* (Los Angeles: Semiotext[e], 2011), 20.

12   Schmitz, "Atmospheric Spaces," 3.

13   Schmitz, "Atmospheric Spaces," 4.

14   Gins, *Helen Keller or Arakawa*, 186.

15   Jennifer Radden summarizes melancholia's long history of prestige as a symptom of genius and a source of celebrity—not to mention a performance of hypersensitivity to "atmosphere"—in *The Nature of Melancholy* (London: Oxford, 2000), 12–19. On women's exclusion from this history, see Juliana Schiesari, *The Gendering of Melancholia* (Ithaca, NY, and London: Cornell University Press, 2002).

16   James Berger, "Testing Literature: Helen Keller and Richard Powers's Implementation of [H]elen," *Arizona Quarterly* 58 (2002): 109–37. https://doi.org/10.1353/arq.2002.0020.

17 For a survey of these photographs and Keller's hypervisible intimacy with the auto-mated communication technologies that exemplified enlightened modernity, with a nod to Benjamin's aura, see Fuss, *The Sense of an Interior*, 135–37. On their disin-genuousness and ambivalent exploitation of Victorian sentimental culture, see Mary Klages, *Woeful Afflictions: Disability and Sentimentality in Victorian America* (Philadelphia: University of Pennsylvania Press, 1999),

18 Braille reading praxis unfolds in a kind of flight pattern, with most fingers actually lifted into the air while fingers in contact with the page move constantly. See Marta L. Werner, "Helen Keller and Anne Sullivan: Writing Otherwise," *Interval[le]s* 2.2–3.1 (2008–9): 958–96. https://doi.org/10.2979/tex.2010.5.1.1.

19 Vilém Flusser, *Towards a Philosophy of Photography* (London: Reaktion, 1983), 36, 8.

20 Keller, *World*, xii.

21 On Keller's merging with her 'others,' see Fuss, *Sense of an Interior*, 109–10; and on the second self that united Keller with such raced others as W.E.B. Du Bois, Fretwell, 221–56.

22 Keller, *World*, 49.

23 Elroy Avery, *School Physics* (New York and Chicago: Sheldon, 1896), 292.

24 Keller, *Teacher*, 8, 38.

25 Keller, "My Story," *The Youth's Companion* 3476 (1894): 3.

26 Keller, *The Story of My Life* [1903], ed. James Berger (New York: Modern Library, 2004), 20.

27 Keller, *Story of My Life*, 83.

28 Jim Swan, "Touching Words: Helen Keller, Plagiarism, Authorship," in *The Construction of Authorship: Textual Approaches to Law and Literature*, ed. Martha Woodmansee and Peter Jaszi (Durham, NC: Duke University Press, 1994), 57–100.

29 Gernot Böhme, *The Aesthetics of Atmospheres*, ed. Jean-Paul Thibauld (London: Routledge, 2016), 1.

30 Keller, *Midstream*, 286.

31 Keller, *Story of My Life*, 40.

32 Keller, *My Religion* (New York: Doubleday, 1927), 2.

33 Emanuel Swedenborg, *Other Planets* [1758], trans. George F. Dole (West Chester, PA: Swedenborg Foundation), 49–50.

34 Keller, *World*, 101.

35 Keller, *World*, 126.

36 Keller, *World*, 111.

37 Keller, *World*, 10.

38 Keller, *Midstream*, xxi.

39 Keller, *World*, 4.

40 Keller, *Midstream*, 484.

41 Fuss, *Sense of an Interior*, 146.

42 Fuss, *Sense of an Interior*, 147.

43 Gins, *Helen Keller or Arakawa*, 238.

44 Gins, *Helen Keller or Arakawa*, 77.

45 Gins, *Helen Keller or Arakawa*, 93.

46 Gins, *Helen Keller or Arakawa*, 166.

47 Gins, *Helen Keller or Arakawa*, 163.

48 Gins, *Helen Keller or Arakawa*, 99.

49 Braddy, Foreword, xx.
50 Jade Stewart, *Revolutions in Air; A Guidebook to Smell* (New York: Penguin, 2021), 3–6.
51 Keller, *World*, 44.
52 Keller, *World*, 60.
53 Keller, *World*, 74.
54 Georg Simmel, "Sociology of the Senses" [1909], in *Sociology. Inquiries into the Construction of Social Form*, trans. and ed. A.J. Blassi, A.C. Jacobs, and M. Kajirathinkal, 2 vols (Leiden: Brill, 2009), vol. 2, 577–78. See also Barbara Carnevali, "Simmel, the Senses, Social Sensibility, and the Aesthetics of Recognition," *Simmel Studies* 21 (2017), 9–143. https://doi.org/10.7202/1043789ar.
55 Simmel, "Philosophy of Landscape" [1912], in *Theory, Culture and Society* 24 (2007): 26.
56 Heather Keenleyside, *Animals and Other People: Literary Form and Living Beings in the Long Eighteenth Century* (Philadelphia: University of Pennsylvania Press, 2016), 23.
57 Keller, *World*, 43.
58 Keller, *World*, 65.
59 Keller, *World*, 44.
60 Keller, *World*, 67.
61 Keller, *World*, 68.
62 Keller, *World*, 69.
63 Keller, *World*, 69.
64 Keller, *World*, 46.
65 Keller, *World*, 71.
66 Keller, *World*, 46.
67 Keller, *Midstream*, 48.
68 Emanuele Coccia, *The Life of Plants; A Metaphysics of Mixture*, trans. Dylan J. Montanari (Cambridge: Polity, 2018), 29.
69 Coccia, *Life of Plants*, 29.
70 Keller, *World*, 77.
71 Keller, *Sense and Sensibility*, *The Century Magazine* 77 (February 1908): 576.
72 Keller, *Sense and Sensibility*, 27.

10

Questions of Visibility: Aerial Relations across Society and the Environment, as Revealed by COVID-19

Savannah Schaufler

As I sit in front of my screen, writing this text in the summer of 2022, I find myself reminiscing about air, inequalities, and the ongoing COVID-19 pandemic—a pandemic that swept across the entire globe in just a few weeks at the beginning of 2020, causing the world organizations to spiral out of control, and leaving the population both figuratively and literally breathless. Since the first cases of "pneumonia of unknown cause" were reported to the World Health Organization (WHO) in December 2019 in the city of Wuhan in China, the new virus spread worldwide abruptly and with such immense force that humanity was blindsided by it.[1] For two years now, humanity has been living in a pandemic that seems to have no end. Although a large part of the world's population is no longer in lockdown, the effects of this human-induced environmental outcry are even more visible. The virus has triggered not only fear of the unknown, but also serious consequences for health, the psyche of individuals, social interaction, and economy.[2] Recognizing that COVID-19 will not be eradicated and is more likely to become endemic in the future, this global phenomenon has revealed the existing exploitation and degradation of Earth and the unequal distribution of natural resources.[3]

Air is composed of a multitude of gaseous, solid, and liquid substances that occupy mass and space.[4] As such, air is matter consisting of many small particles, such as dust, soot, and droplets.[5] To trace the emergence of this relatively new virus and to recognize the intertwined relationships between humans, the more-than-humans, and the environment, air can be used as a metaphor. This allows us to illustrate the spiderweb of human-environmental

Savannah Schaufler, "Questions of Visibility: Aerial Relations across Society and the Environment, as Revealed by COVID-19" in: *Imagining Air: Cultural Axiology and the Politics of Invisibility*. University of Exeter Press (2023). © Savannah Schaufler. DOI: 10.47788/WSKQ2452 linkages and illustrating the extremely "complex family of interconnected problems."[6] The attempt to consider air as material and the dissolution of the living and the material allows visualizing air in its fluidity. In recognition of Jane Bennett's discourse on "vital materiality," this materiality (things, objects, items, etc.) can no longer be viewed as something passive, controllable, or constructible by human beings, but instead presents objects as lively. With the foregrounding of the material in Bennett's work, alongside a broader turn toward the material across the humanities in recent decades, the complexity of materiality as an actant in human and more-than-human relationships is presented.[7] Drawing toward an increasing emphasis on the need to move beyond the centrality of the human subject and addressing how human cultures are part of many other material cultures in the world, the very complex presence of air that presents itself as an agency of matter is recognized. This enables a reappraisal of air's value towards an axiological framework, bringing to the discussion a renewed and differentiated aerial relation of COVID-19. In the context of air, its relationship to the intertwined structural, social, and biological upheaval caused by the COVID-19 pandemic can be analyzed. However, once one addresses one aspect of an issue, other aspects emerge through discourse and thematic exploration, leading to even more branching pathways that are difficult to sparse.

Following the worldwide spread of COVID-19 in the air, this chapter discusses and relates to various forms of aerial harm. This allows air and the figurative notion of the dispersal of COVID-19 through aerosols for the materialization of objects and visualization of their interconnectivity. This chapter presents the peculiarities of the relationship between air and COVID-19 through its historical and evolutionary aspects as well as theories of human conduct. In thinking through these encounters using the lens of COVID-19, the intertwined ecological, structural, and societal flaws can be viewed in different ways. This illustration of the velocity and pervasiveness of the problems of exploitation, visibility, and inequality resulting from the human mode of production is framed in the discussion through an evolutionary lens. Taking this interdisciplinary approach and incorporating multiple thematic perspectives provides an opportunity to address issues of the invisible unseen, as all human, physical, and environmental aspects are intimately connected. In particular, these aspects of air's materiality and its usage as a metaphor allow us to describe the connections between humans, nonhumans, and the environment, as the element of air itself is life-giving to all living and nonliving organisms.

The Rise of Social and Environmental Inequality

Attempting to look at the evolutionary history of COVID-19 points us towards the unequal interdependence between humans, nonhumans, and the environment. Tracing these roots of inequality, Heather Pringle argues that they lie in human ancestors—in hunter-gatherer societies.[8] Throughout evolution and the "increasing complexity of the human brain," hunter-gatherer societies were able to develop new working conditions and access new natural resources.[9] New economic forms, such as agriculture, animal farming, and gradual sedentarization emerged 11,700 years ago during the Holocene epoch and are positioned as part of the Neolithic Revolution.[10] Although most of these societies were egalitarian, factors such as population pressure, spatially concentrated resources, production controls, and the manipulation of social networks may have created inequalities in these societies as well.[11] Egalitarian societies refer to social groups and communities in which, in principle, all members have equal access to key resources (food, goods, land, etc.) and no member can permanently exercise power over others.[12] The social status of individuals in egalitarian groups depends primarily on their abilities and will, with political and social equality prevailing. The cohabiting societies subsequently became larger and larger, which led to growing and flourishing cities, resulting in a new economic system and leading to the "intensification of social hierarchies."[13] This radical change in societies resulted not only in the rise of inequalities, but also in humans increasingly altering and disrupting the environment by intruding into functioning ecological systems.[14]

Consequently, inequalities can be divided into two types—those of social relations and those between humans and the environment. There are many aspects to social inequality as it problematizes the unequal distribution of society's goods and resources. According to sociologist Ruth Levitas and colleagues, resources are defined as social, material, and economic, including access to both private and public services.[15] Looking at history, social inequality has existed throughout human evolution and is experienced along socially defined categories.[16] Inequality can be based on ethnicity, gender, age, sexual orientation, religion, and the uneven distribution of power.[17] These inequalities arise at the intersections of different categories of discrimination against a person or a group of individuals, which is referred to as intersectionality—this term was coined by the American lawyer Kimberlé Crenshaw in the 1980s to express the intersection of discrimination categories.[18]

The second inequality results from the unequal relationship between humans and nature. Since nature provides use of value for human life, humans increasingly altered and shaped the environment with the onset of sedentarization.[19] Natural products are used in excess, with wood, for example, being used to produce energy and houses, and the reacquired land area is used for cultivating monocultures and the production of food.[20] This overproduction leads to a circle of overexploitation of natural resources, resulting in a disequilibrium between humans and the environment.

Human intrusion into the natural world is destabilizing the environment and causing microorganisms to find new hosts, resulting in an increasing occurrence of new contagious diseases and infections.[21] The emergence of new diseases results from the fact that microorganisms are not only found in nature, but also live on and in humans.[22] However, while microbes can

live in symbiosis or commensalism, they can also be threatening and pathogenic to humans and other living and nonliving organisms.[23] In this context,
Jane Bennett describes the concept of "vital materiality," where she argues
that human beings are unaware "of the ocean of microbes that is both inside
and outside our bodies."[24] Bennett's theory is based on the idea that bodies
interact with each other, resulting in the formation of a collective of disparate
elements. She describes this collective as an "assemblage," citing the example
of the curvature of the elbow, which forms an ecosystem for a majority of
different strains of bacteria.[25] These bacteria permanently coat, hydrate, and
protect the skin by breaking down the fats that the skin produces.[26] Given
that microorganisms interact with each other, as humans do, they can subvert
evolutionary pathways, thus enabling the spread of new infectious diseases.[27]
Evolving with changing times are anthropogenic activities that lead to
zoonotic infections—urbanization, extractive agriculture and land use, extensive human travel, and global export and import of commodities such as
fruits, vegetables, meats, and plants—leading to increasingly frequent disease
outbreaks.

This intervention, accelerated by the resulting more intimate relationship
between humans and animals, has led to the increased emergence and spread
of new infectious diseases caused by pathogenic viruses and bacteria, as well
as the increase of zoonotic diseases spread through the air.[28] Using Dipesh
Chakrabarty's words, "the pandemic connotes a time when our recognition
of the microbial world we live amid can no longer be contained."[29] That
being the case, the relatively new infectious disease COVID-19 represents
a direct response to this interweaving of worldly relationships.

By inhaling infectious aerosols or droplets smaller than 5 μm, the infectious disease—COVID-19—is transmitted from human to human.[30] During
this process, transmission usually occurs through loud breathing, talking,
coughing, sneezing, or singing, which causes salivary droplets containing
virus to be propelled into the surrounding air, which in turn are inhaled by
other individuals, who may thereby deposit potential pathogens in their
respiratory tract.[31] As coughing and sneezing are aerosol-producing actions,
they are triggered by a stimulus or reflex and ejected into the air by involuntary expulsion of respiratory and salivary fluids from the mouth and
nose.[32] Historically, sneezing has been discussed as both a positive and
negative sign.[33] Even Hippocrates emphasized that "sneezing is dangerous
only prior to or following a pulmonary disease."[34] In fourteenth-century
Europe, during the outbreak of the Black Death, caused by the bacterium
Yersinia pestis, the prayer "May God bless you," declared by Pope Gregory,
was desired after each sneeze to protect against possible infection with the
plague.[35] While there have been numerous epidemic outbreaks throughout
history (e.g., the Plague of Justinian, the Black Death, numerous outbreaks
of cholera and influenza, SARS, and MERS) owing to unknown pathogens,
the COVID-19 pandemic nevertheless made social and structural abuses
even more apparent.[36]

The Essence of Air

The dispersion of aerosols in the air allows the air to be recognized as an archive, enabling the connections of aerial relations. Describing the various reflections of air realities and narratives transported and mediated in the air provides the opportunity, as the transdisciplinary researcher on politics of air Marijn Nieuwenhuis writes, to "[think] through the air."[37] As air travels across global territories, it contains different realities of life, natural processes, scientific knowledge, and human politics.[38] To this end, theorist Eva Horn writes, air represents a "medium of life."[39] In doing so, she argues from an epistemological perspective that the subjectivity of air should not be forgotten as air is a sustaining element of humanity. Similarly, the alchemist, philosopher, and chemist Robert Boyle engaged with the ideas and realities of air, albeit in a more chemical context. He assumed "that there may be in the [a]ir some yet more latent [q]ualities or [p]ower ... [f]or this is not ... a [s]imple and [e]lementary [b]ody, but a confus'd [a]ggregate of effluviums from such differing [b]odies."[40] Boyle argues that air is a large heterogeneous element, an aggregation consisting of atoms and molecules forming a large association, inhabiting different stories of life itself.

In particular, the global pandemic triggered by COVID-19 (SARS-CoV-2) demonstrates that air is not a simple element that living organisms need to breathe, but reflects sociopolitical, biopolitical, and material realities and interactions.[41] To illustrate these interactions, the question of how "air enters into human and geographic life" is explored through drawing on research from scholars such as Timothy Choy, Peter Adey, and Magdalen Górska to discuss the great "ocean of air."[42] As air contains a vast amount of feelings, violations, and senses, human beings most often forget to recognize its essentiality to life.

In his text on Beijing's air and air pollution, "Air's Substantiations," Timothy Choy shows the great power that air can have on human life. In doing so, he outlines the affective and aesthetic dimensions that air and the atmosphere inherently possess.[43] Drawing on his lived experience of how air pollution affected and impacted his own health, Choy indicates that humans are never completely separated from toxins and their affective effects suspended in the air.[44] Exposure to chemicals and respirable dust thus becomes an anticipatory vehicle, showing that atmospheric variability can act as a harbinger of climate disasters such as hurricanes and storms.[45] While pointing to the materiality, histories, and lived experiences in the air, this description illustrates air "as an object of concern, care, and investment among those affected by it," as Kaushik Sunder Rajan writes in the introduction to Choy's chapter.[46] Choy argues that though air is fascinating through its invisibility, its materiality should not be neglected, and climate and air can affect people in many ways, raising the social, geo- and biopolitical aspects with which air is interwoven.[47]

Looking more closely at this violation and abuse of air, the marginalization and exclusions that are transported through air become visible. Air is precisely a figuration embedded in many social, biopolitical, and economic perspectives.[48] This is also evident through Peter Adey's consideration of the "Air/Atmospheres of Megacities."[49] The effective power of air to bring human inequalities into focus is illustrated by the so-called megacities, where mostly lower-income populations are more affected by elevated levels of air pollutants and bad air quality.[50] In this context, a megacity is defined as an agglomeration area that usually has more than 10 million inhabitants.[51] These include, for example, New York City, Mexico City, Rio de Janeiro, Tokyo, Mumbai, London, and Paris. Adey views megacities through the air of difference and argues that "[a]n analysis of air reveals who belongs and who does not, who is deserving and who is not in a constellation of megacity inequality."[52]

In this sense, anthropologist Celia Lowe understands the sky as a biopolitical zone or cloud that covers the sky and contains pathogenic realities.[53] In her argument, she takes the H5N1 (avian flu) outbreak in Indonesia as a starting point.[54] As with Jane Bennett, she also foregrounds the microbiological world and the disregard for constant exchange between human and microbial environments.[55] In doing so, she coins the concept of the "biopolitical cloud" as a process beyond human control.[56] Air, although often abused and altered by human activities, can therefore be considered as an entity, containing dangerous and deadly particles. A breath, an exhalation, once it is released into ambient air disperses aerosols from one place to another, from one exhalation to another inhalation.

Against this backdrop, the COVID-19 pandemic can be understood as a biopolitical cloud, floating in the airspace, revealing, as Adey shows, who is worth protecting and who is not. For this, the concept of "suffocation" in air defined by cultural studies and feminist philosopher Magdalena Górska may help to explain the transported, multiple narratives through and within the air itself.[57] Górska attempts to describe the aspect of suffocation on "external and internal borders" using panic attacks and the underlying politics of social norms.[58] Similarly, African diasporic researcher Gabriel O. Apata discusses asphyxiation in air.[59] In particular, he addresses underlying racism and its prolonged invisible process, as history shows.[60] He explains how the air—the social air—through its systemic and structural racisms restricts the air communities of Black and Indigenous People of Color (BIPOC) breathe, diluting it more and more until there is no breathable air left.[61] However, in this chapter, this suffocation refers to polluted air within the body itself and the environment in which the body lives, produced by the expulsion of exhaust fumes, the spread of viruses. Transporting not only harmful particles for humans, more-than-humans, and the environment, but also social and biopolitical narratives of marginalization, air becomes a vehicle creating a sense of suffocation in an ambivalent environmental, social, and biopolitical climate.

Considering the biopolitical cloud and the sense of suffocating in air, the COVID-19 pandemic resembles an abuse of air and planetary health that has been brought about by human environmental harm. The cloud, transported in the air, transmits infections, danger, concern, and fear for those affected and asphyxiating in air. Representing this biopolitical cloud, the oxygen inhaled through breath, these pathogen particles and narratives are respired as well. Following these dispersed particles of dirt and pathogenic microorganisms suspended in the air, a transformation of not forgetting, of recognizing air's materiality, the biopolitical and sociopolitical entanglements of humans and all other living and nonliving organisms inhabiting the environment, unfolds.

Air Pollution and COVID-19

Continuing the thought of polluted and pathogenic air and clouds, a connection between the COVID-19 pandemic and air pollution unravels. Air pollution is the result of a multitude of releases of gases, solids (dust, pollen), and liquids, fossil fuels.[62] These factors, once released into air, primarily contribute to changes in its chemical composition, resulting in continuous pollution of ambient air.[63] Increasingly, greenhouse gases such as carbon dioxide or methane and various aerosols including sulfates are released into air.[64] This invisible pollution has far-reaching consequences, and poses a growing threat to living and nonliving organisms and the environment.

The origins of air pollution date back to the time of the Greco-Roman Empire.[65] Most notably, James Watt's improved steam engine of 1784 is credited as the starting point for the rising causes of air pollution. Historian William M. Cavert shows how London, plagued by coal smoke from factories and smokestacks, heated up the environment.[66] This persistent filth not only led to severe health problems, but also became a figurative description of life in seventeenth-century London. Venturing through historical events, the experiences convoluted by air echo a reservoir of lived realities.

Following the trail of realities echoed in the harmful particles swirling around in air, the COVID-19 pandemic triggered humans' consciousness of the physical spread of air and its harmful containments. Philosopher Timothy Morton's concept of "hyperobject" appears fitting for conceptualizing the connection between air pollution, COVID-19, and its aerial relations.[67] Morton argues that hyperobjects are "vicious," "nonlocal," and "occupy a high-dimensional phase space that results in their being invisible to humans for stretches of time."[68] These phenomena are of expansive entities that stretch across time and space while in indirect mundane contact, and thereby have considerable influence on human and more-than-human interaction.[69] Among these Morton includes the solar system, the biosphere, and "long-lasting product[s] of direct human manufacture."[70] Drawing on the poisoned nature of these hyperobjects, the ambivalent ways in which air pollution and COVID-19 operate on a human and ecological level is revealed. As hyperobjects interact not only through "objects," but also through narratives, such as cultural and historical memory,[71] the boundaries between "human existence" and nature blur, highlighting their multiple connections with nature and prevailing (economic) systems.[72]

As Robert-Jan Wille, a scholar of history of climate science and atmospheres, and Kurram Shehzad, Muddassar Sarfraz, and Syed Gulam Meran Shah, who specialize in chemistry and business management, point out, the atmosphere and air are linked in multiple ways to the pandemic triggered by COVID-19.[73] As a result of the restrictions imposed by the pandemic, such as closures, travel bans, traffic restrictions, social distancing, and less business activity, there was a decrease in air pollution over many (mega) cities such as Los Angeles, Delhi—where the India Gate could be seen—and Manila, from which the Sierra Madre Mountains were visible again.[74] These examples show that the possibility of living under blue skies and in clean air is not a dream and would come about if emissions were reduced and fossil fuels were no longer burnt.[75] This idea, however, would only come to fruition if the entire world population were in a strict lockdown. The power that ideas and ways of thinking can have in societies is described by political scientist Maarten Hajer and Dutch writer Wytske Versteeg in their elaboration of "imagining the post-fossil city."[76] They use this example to discuss the importance that imaginaries, stories, and images can have on societies: A lack of positive imagination creates a paralyzing constraint that suppresses improvement.[77] This lack of imagination, according to the authors, results from modern economies not being able to sustain life itself without the use of fossil fuels.[78] This usage and this dependence are not only interwoven in everyday routines, but are also perpetuated in societal values.[79] Living under blue skies has been scarce, only caused by lockdowns during which economies and societies have been forced to reduce fossil fuel emissions.

In this context, and considering the problems that are rapidly approaching societies, a current new "worldview" of forgetting, of the invisible, of the tacit, is present. Thereby, I argue that air pollution can be understood as a kind of pandemic, since it not only transports pathogens across borders, nations, and cultures, but also conveys narratives of exclusion, of survival of the fittest. This silencing and turning away from secular problems is strongly embedded in values and everyday routines. These new "worldviews" make it possible to decimate cognitive dissonance. Acceptance and perception of things—air pollution—can change individual behaviors and actions according to their conformity.

The theory of cognitive dissonance was established by Leon Festinger (1957) to explain conflicting attitudes and behaviors of individuals. The theory states that there are two or more "cognition pairs or knowledge elements."[80] Each individual strives for consistency in their beliefs.[81] Therefore, if an individual has two cognitions—attitudes, values, norms, beliefs, and/or emotions—that are inconsistent with one another, the individual attempts to reduce this tension, called cognitive dissonance.[82] It is an individual's reality that determines the content. If reality exerts pressure on an individual's cognition, an attempt is made to establish a consonance—a balance—even if in the process a new reality is created in which the individual's justification is linear.[83] As Festinger writes, once this "reality ... impinges ... [it] will exert pressures in the direction of bringing the appropriate cognitive elements into correspondence with that reality."[84]

The mind game of defining air pollution as a pandemic clouds the conceptions of living under blue skies, which is why the improvement of global air quality was short lived during the lockdown and the average decline varied between countries.[85] Regarding this transient phenomenon, studies by spatial ecologist Zander Samuel Venter and colleagues show that a dramatic decline in economic activity and travel accounted for the decline in global pollutant emissions, including concentrations of sulfur dioxide (SO_2), nitrogen dioxide (NO_2), ozone (O_3), and particulate matter ($PM_{2.5}$).[86] The resulting negative feedback is also reported, as there is an implicit correlation between areas of high air pollution and COVID-19 deaths.[87] A 2020 study by four scientists at Harvard University's T.H. Chan School of Public Health presented findings indicating that polluted air increases COVID-19's lethality.[88] They analyzed data on $PM_{2.5}$ concentrations in ambient air and COVID-19 deaths in several US counties.[89] They examined the association of impairment from long-term $PM_{2.5}$ exposure with an increased mortality risk among COVID-19 patients. According to these results, long-term pollution exposure to $PM_{2.5}$ was associated with a 15 percent increase in COVID-19 mortality rate in the areas studied.[90] These data again underline the convoluted relationship that allows us to view air pollution as a pandemic—contaminated not only by pollutant emissions, but also by viral pathogens.

Air is often perceived as merely invisible owing to the worldview of forgetting; "the air we inhale commonsensically appears to be *just* air."[91] Linking these thoughts, it becomes apparent that air imaginaries are of an extremely complex nature, involving "a hybrid human-non-human entanglement which nevertheless has very material, socially uneven, consequences."[92]

"*Just* Air"

As air travels and floats across many frontiers and terrains, noxious particles and narratives move within it. Air's composition changes are influenced by emissions of pollutants, breath, stories, and tears, containing numerous narratives. Using the term "*just* air," borrowed from the interdisciplinary working geographers Anneleen Kenis and Maarten Loopmans (2022), this section specifically grapples with the meaning of the word "*just*."[93] This word has many connotations, but it is the meaning of justice that steps into focus in the context of air. On one hand, the word undermines the vital function of air because of its invisibility and because air is not "*just*" air.[94] On the other hand, it emphasizes the value of air and the unequal access to clean air, because air is not "*just*." Air is not "*just*" nature, but a space where biopolitical and sociopolitical conflicts take place.[95] This problem of unjust air and suffocation in air is reflected in racism and discrimination debates. In particular, in the United States, a strong socially rooted racism hovers in the air. Lizzie Wade, a science writer focusing on archeology and anthropology, believes that "[e]very ancient society studied was much more equal than the United States is today."[96]

While the entire world was struggling with the COVID-19 pandemic, yet another racially motivated murder by the police occurred in the United States. An unarmed Black man named George Floyd was killed by a police officer during a violent arrest on May 25, 2020, in Minneapolis, Minnesota: Floyd's neck was pressed to the ground for over eight minutes, preventing him from breathing.[97] A video of the incident sparked worldwide protests against police violence and racism. In this context, the global fight against and containment of the spread of COVID-19 and the global outcry of protest movements seem to have ambivalent relationships. Although social distancing is one of the key behavioral rules for halting further spread of COVID-19, Black Lives Matter (BLM) protests around the world mobilized to fight and stand up for social justice in public.[98] This unexpected end to social distancing measures for the period of the protests called attention to the deep-rooted racial inequalities toward African-Americans in health care.[99] Many argued in favor of the protests because protests themselves are essential to public health, as the mortality rate among American BIPOC is twice that of white Americans. The protests symbolized the fight for justice, but also provided a "portal" of change, an opportunity to fight back, amid a pandemic that was killing Black people.[100]

The BLM movement was born out of the acquittal of George Zimmerman, a white Florida neighborhood watch volunteer who shot and killed 17-year-old African-American Trayvon Martin. The movement was initiated by three women of color, Alicia Garza, Opal Tometi, and Patrisse Cullors.[101] In response to this act and many others (such as the 2014 killings of Eric Garner and Michael Brown, and of Tamir Rice and Javier Ellis in 2020), the BLM movement emerged to fight against structural racism and "to intervene in violence inflicted on Black communities by the state and vigilantes. By combating and countering acts of violence, creating space for Black imagination and innovation, and centering Black joy."[102]

Narratives of structural violence, racism, and discrimination transported across historical airways denote meaning-making narratives that not only influence how the environment is perceived, but also represent established narratives that are invested with legitimacy.[103] After all, narratives are not merely stories but are patterns of narration that, contribute to whether stories are believed or not, regardless of their content.[104] Narratives can therefore reflect facts, as well as obscure or twist realities by subjectively telling how the world is perceived.[105] Most notably in the United States, narratives about predominantly Black men with low socioeconomic status are carried throughout the entirety of America's origin story, when enslaved African people were first brought to the New World.[106] Following the eighteenth- and nineteenth-century belief that blacks were "mentally inferior, physically and culturally inferior," images and stereotypes of violent, criminal, and impoverished Black men continued to plague the Black community. [107]

American sociologists and race theorists Michael Omi and Howard Winant attempt to explain the normalization of these stereotypes and their acceptance as truth with their concept of "colorblind racism."[108] By introducing the concept of "colorblind racism," they theorize the historical emergence of racial stereotypes and structural racism.[109] With the passage of civil rights reforms in the mid-1960s and the civil rights movement in the United States, a "democratizing, inclusionist, and egalitarian" narrative emerged.[110] Colorblind racism appears as an egalitarian narrative that claims all people are equal.[111] In this, "colorblind racism" resembles a "race-neutral" reinterpretation.[112] Thus, racism can affect anyone, regardless of skin color, gender, or socioeconomic status. The problem with this is the forgetting and deliberate negation, structural legacies of slavery, and the constant denigration and exclusion.[113] This dangerous liberal ideology provides a false sense of security to those who do not face racial oppression in their daily lives.[114] This ambivalence and created reality, of the everyone, an equal approach, can also be reflected in the cognitive dissonance theory and the worldview of forgetting. If two or more cognitions oppose each other, one tries to approximate them by creating one's own reality.

The creation of these new realities, as they relate to the past and to the truth about the marginalization of groups because of their skin color, gender, or socio-economic status, suffocate and dissolve in air, in individuals' minds. This suffocation in air is therefore not just a metaphor; it becomes the reality of many. Accordingly, air is not "*just* air," but it illustrates the biopolitical harm to which those oppressed by structural powers are exposed.

Conclusion

Viewing the entanglements that the metaphorical way of looking at air allows, the complexity of earthly interconnections manifests itself. Focusing on one issue of concern without thinking of or linking it to another issue is almost impossible. The vision and understanding of how interconnected economies, nation-states, health, inequalities, social and environmental issues, and representations really are has been recognized through the problems that individuals, societies, and states have faced and continue to face. These problems in modern societies are intricately intertwined and confounded, making it difficult to detangle them from their surroundings and engage with them. The approach I have taken in this chapter is to revalue air and its materiality by discussing, challenging, and reflecting on current crises and scholarly conversations as I have engaged with air through COVID-19. Through this, this chapter has attempted to address and unwind the entanglement of worldly complexity to move beyond the synthesis of aerial dialogues and discussions.

The airborne relationships between society and the environment revealed by the COVID-19 pandemic may provide a basis for disentangling issues of environmental and societal damage. Therefore, by looking at this through the lens of evolution and using theories of human behavior—without justifying actions and attitudes—explanations can be created that serve as an aid in working toward meaningful societal change. As the Indian writer Arundhati Roy argues, the COVID-19 pandemic can be understood as a "portal."[115] This is not a portal that allows a return to normality, but a new vision, a new "worldview," that allows for a more nuanced understanding of the interdependencies between humans, more-than-humans, and the environment. The question of whether the COVID-19 pandemic is truly a "portal," to borrow Roy's term, remains to be answered. Crises can spark a debate about making the invisible visible, which can bring about a change in representation and societies.[116] Every crisis draws attention to the invisible and attempts to make it visible. The sordid tales of the enacted through worldviews of forgetting and the exploitation of natural resources humanize the invisible and show the immense impact and dangers of exploitative activities. Fighting injustice and environmental degradation is an ongoing struggle, prompted by crises such as the COVID-19 pandemic.

Luce Irigaray claims that a "forgetting of air" should not arise, for every living being, every plant, every organism lives in the air.[117] Air is not only an element; it transports—across borders—a multitude of political, cultural, and social realities floating above all earthly things containing concepts of ambivalence.[118] The materiality of air allows us to feel and sense the intimacy between the problems of the world—the social and environmental damage. Through the concept of aerial relations and air's materiality, air can be used to discuss entanglements, as air is not a neutral space but rather the essence of life that is significantly polluted through fear, anger, exploitation, tears, and hardships. Once released into air, these pollutants cannot be taken back.

Notes

1   World Health Organization (WHO), "COVID-19-China," January 5, 2020, https://www.who.int/emergencies/disease-outbreak-news/item/2020-DON229, accessed May 15, 2022.

2   Dominika Maison, Diana Jaworska, Dominika Adamczyk, and Daria Affeltowicz, "The Challenges Arising from the COVID-19 Pandemic and the Way People Deal with Them. A Qualitative Longitudinal Study," *PLoS ONE* 16, no. 10 (October 11, 2021): 1–17. https://doi.org/10.1371/journal.pone.0258133

3   Nicky Phillips, "The Coronavirus Is Here to Stay—Here's What That Means," *Nature* 590 (February 16, 2021): 382. https://doi.org/10.1038/d41586-021-00396-2

4   Detlev Möller, *Luft: Chemie, Physik, Biologie, Reinhaltung, Recht* (Berlin: De Gruyter, 2003), 1–2.

5   Möller, *Luft*, 1–2.

6   Dipesh Chakrabarty, "The Politics of Climate Change Is More than the Politics of
     Capitalism," *Theory, Culture & Society* 34, no. 2–3 (2017): 25–37. https://doi.
     org/10.1177/0263276417690236; Eva Horn, "The Case of Air," Anthropogenic
     Markers, April 22, 2022, https://www.anthropocene-curriculum.org/anthropogenic-
     markers/critical-environments/contribution/the-case-of-air, accessed April 27, 2022.
7   Christopher N. Gamble, Joshua S. Hanan, and Thomas Nail, "What Is New
     Materialism?," *Angelaki* 24, no. 6 (November 2, 2019): 111–34. https://doi.org/
     10.1080/0969725X.2019.1684704
8   Heather Pringle, "The Ancient Roots of the 1%," *Science* 344, no. 6186 (May 23,
     2014): 822–25. https://doi.org/10.1126/science.344.6186.822
9   Oded Galor, "The Journey of Humanity: Roots of Inequality in the Wealth of
     Nations," *Economics and Business Review* 6 (20), no. 2 (2020): 7–18. https://doi.
     org/10.18559/ebr.2020.2.2
10   Galor, "The Journey of Humanity," 13.
11   Hillard S. Kaplan, Paul L. Hooper, and Michael Gurven, "The Evolutionary and
     Ecological Roots of Human Social Organization," *Philosophical Transactions of the
     Royal Society B: Biological Sciences* 364, no. 1533 (November 12, 2009): 3289–99.
     https://doi.org/10.1098/rstb.2009.0115; Eric Alden Smith and Brian F. Codding,
     "Ecological Variation and Institutionalized Inequality in Hunter-Gatherer Societies,"
     Proceedings of the National Academy of Sciences 118, no. 13 (March 30, 2021): 1–9.
     https://doi.org/10.1073/pnas.2016134118
12   Pringle, "The Ancient Roots of the 1%," 822.
13   Lizzie Wade, "Livestock Drove Ancient Old World Inequality," *Science* 358, no. 6365
     (November 17, 2017): 850. https://doi.org/10.1126/science.358.6365.850
14   Antonio Cascio, Mile Bosilkovski, Alfonso J. Rodriguez-Morales, and Georgios
     Pappas. "The Socio-Ecology of Zoonotic Infections," *Clinical Microbiology and
     Infection* 17, no. 3 (March 2011): 336–42. https://doi.org/10.1111/j.1469-
     0691.2010.03451.x
15   Ruth Levitas, Christina Pantazis, Eldin Fahmy, Davird Gordon, Eva Lloyd, and
     Demi Patsios, *The Multidimensional Analysis of Social Exclusion* (Bristol: University
     of Bristol, 2007).
16   David B. Grusky, "The Past, Present, and Future of Social Inequality," in *Social
     Stratification* (London: Routledge, 2001), 3–51. https://doi.org/10.1016/B0-08-
     043076-7/01974-4
17   Louise Warwick-Booth, *Social Inequality* (London: SAGE Publications, 2018),
     2–26.
18   Warwick-Booth, *Social Inequality*, 2–26; Kimberlé Crenshaw, "Mapping the Margins:
     Intersectionality, Identity Politics, and Violence against Women of Color," *Stanford
     Law Review* 43, no. 6 (July 2022): 1241–99. https://doi.org/10.2307/1229039
19   Thomas Kirchhoff, "Zum Verhältnis von Mensch und Natur," Bundeszentrale für
     politische Bildung, March 16, 2020, https://www.bpb.de/shop/zeitschriften/
     apuz/305897/zum-verhaeltnis-von-mensch-und-natur/#footnote-target-8, accessed
     July 20, 2022.
20   Kirchhoff, "Verhältnis von Mensch und Natur," Konrad Ott, Jan Dierks, and Lieske
     Voget-Kleschin, *Handbuch Umweltethik* (Stuttgart: J.B. Metzler Verlag, 2016), 2–4.
21   Institute of Medicine (US) Committee on Emerging Microbial Threats to Health
     in the 21st Century, *Microbial Threats to Health: Emergence, Detection, and Response*,

ed. M.S. Smolinski, M.A. Hamburg, and J. Lederberg (Washington, DC: National Academies Press, 2003). https://doi.org/10.17226/10636

22 Institute of Medicine (US), *Microbial Threats to Health*.

23 Julian Davies, "Microbes Have the Last Word: A Drastic Re-evaluation of Antimicrobial Treatment Is Needed to Overcome the Threat of Antibiotic-resistant Bacteria," *EMBO Reports* 8, no. 7 (July 2007): 616–21. https://doi.org/10.1038/sj.embor.7401022

24 Jane Bennett, *Vibrant Matter: A Political Ecology of Things* (Durham, NC: Duke University Press, 2010), 112–13; Dipesh Chakrabarty, "The Chronopolitics of the Anthropocene: The Pandemic and Our Sense of Time," *Contributions to Indian Sociology* 55, no. 3 (2021): 324–48. https://doi.org/10.1177/00699667211065081

25 Bennett, *Vibrant Matter*, 112–13.

26 Bennett, *Vibrant Matter*, 112–13.

27 Jan Slingenbergh, Marius Gilbert, K. de Balogh, and William Wint, "Ecological Sources of Zoonotic Diseases," *Revue Scientifique et Technique de l'OIE* 23, no. 2 (August 1, 2004): 468–84. https://doi.org/10.20506/rst.23.2.1492; Cascio et al., "Zoonotic Infections," 336–42; Davies, "Microbes Have the Last Word," 611.

28 Levent Akin and Mustafa Gökhan Gözel, "Understanding Dynamics of Pandemics," *Turkish Journal of Medical Science* 50 (2020): 515–19. https://doi.org/10.3906/sag-2004-133; Orhan Kilic, "Pandemics Throughout History and Their Effects on Society Life," in *Reflections on the Pandemic in the Future of the World*, ed. Ali Özer, Cem Korkut, and Muzaffer Şeker (Ankara: Türkiye Bilimler Akademisi, 2020), 13–53. https://doi.org/10.53478/TUBA.2020.026

29 Chakrabarty, "Chronopolitics of the Anthropocene," 333.

30 Cuiyan Wang, Riyu Pan, Xiaoyang Wan, Yilin Tan, Linkang Xu, Cyrus S. Ho, and Roger C. Ho, "Immediate Psychological Responses and Associated Factors during the Initial Stage of the 2019 Coronavirus Disease (COVID-19) Epidemic among the General Population in China," *International Journal of Environmental Research and Public Health* 17, no. 5 (March 6, 2020): 1–25. https://doi.org/10.3390/ijerph17051729

31 Majid Rezaei and Roland R. Netz, "Airborne Virus Transmission via Respiratory Droplets: Effects of Droplet Evaporation and Sedimentation," *Current Opinion in Colloid & Interface Science* 55 (October 2021): 1–12. https://doi.org/10.1016/j.cocis.2021.101471; Nick Wilson, Stephen Corbett, and Euan Tovey, "Airborne Transmission of Covid-19," *BMJ* (August 2020): 1–3. https://doi.org/10.1136/bmj.m3206; Prateek Bahl, Con Doolan, Charitha de Silva, Abrar Ahmad Chughtai, Lydia Bourouiba, and C. Raina MacIntyre, "Airborne or Droplet Precautions for Health Workers Treating Coronavirus Disease 2019?," *The Journal of Infectious Diseases* (April 2020): 1–8. https://doi.org/10.1093/infdis/jiaa189; Santosh K. Das, Jan-e Alam, Salvatore Plumari, and Vincenzo Greco, "Transmission of Airborne Virus through Sneezed and Coughed Droplets," *Physics of Fluids* 32, no. 9 (September 2020): 1–7. https://doi.org/10.1063/5.0022859

32 Murat Songu and Cemal Cingi, "Sneeze Reflex: Facts and Fiction," *Therapeutic Advances in Respiratory Disease* 3, no. 3 (June 2009): 131–41. https://doi.org/10.1177/1753465809340571; Bahl et al., "Airborne or Droplet Precautions," 1–8.

33 Songu and Cingi, "Sneeze Reflex," 131.

34 Songu and Cingi, "Sneeze Reflex," 131.

35   Songu and Cingi, "Sneeze Reflex," 132.

36   Jocelyne Piret and Guy Boivin, "Pandemics Throughout History," *Frontiers in Microbiology* 11 (January 15, 2021): 1–16. https://doi.org/10.3389/fmicb.2020.631736

37   Marijn Nieuwenhuis, "Breathing Materiality: Aerial Violence at a Time of Atmospheric Politics," *Critical Studies on Terrorism* 9, no. 3 (September 2016): 499–521. https://doi.org/10.1080/17539153.2016.1199420

38   Eva Horn, "Air as Medium," *Grey Room* 73 (2018): 8–25. https://doi.org/10.1162/grey_a_00254

39   Horn, "Air as Medium," 8–9.

40   John F. Fulton, "Robert Boyle and His Influence on Thought in the Seventeenth Century," *Isis* 18, no. 1 (1932): 77–102. https://doi.org/10.1086/346688

41   Cymene Howe, "Life Above Earth: An Introduction," *Cultural Anthropology* 30, no. 2 (May 25, 2015): 203–09. https://doi.org/10.14506/ca30.2.03

42   Peter Adey, *Air: Nature and Culture*, Earth Series (London: Reaktion Books, 2014), 121.

43   Timothy Choy, "Air's Substantiations," in *Lively Capital: Biotechnologies, Ethics, and Governance in Global Markets*, ed. Sunder Rajan Kaushik (Durham, NC: Duke University Press, 2012), 121–52.

44   Choy, "Air's Substantiations," 121–52; Timothy Choy, "A Commentary: Breathing Together Now," *Engaging Science, Technology, and Society* 6 (November 10, 2020): 586–90. https://doi.org/10.17351/ests2020.771

45   Choy, "Air's Substantiations," 121–52.

46   Sunder Rajan Kaushik, *Lively Capital: Biotechnologies, Ethics, and Governance in Global Markets* (Durham, NC: Duke University Press, 2012), 1–41.

47   Choy, "Air's Substantiations," 121–52; Choy, "Breathing Together Now," 586–90.

48   Steven Connor, *The Matter of Air: Science and Art of the Ethereal* (Chippenham: Reaktion Books, 2010); Peter Adey, "Air's Affinities: Geopolitics, Chemical Affect and the Force of the Elemental," *Dialogues in Human Geography* 5, no. 1 (March 2015): 54–75. https://doi.org/10.1177/2043820614565871

49   Peter Adey, "Air/Atmospheres of the Megacity," *Theory, Culture & Society* 30, no. 7–8 (2013): 291–308. https://doi.org/10.1177/0263276413501541

50   Anjum Hajat, Charlene Hsia, and Marie S. O'Neill, "Socioeconomic Disparities and Air Pollution Exposure: A Global Review," *Current Environmental Health Reports* 2, no. 4 (December 2015): 440–50. https://doi.org/10.1007/s40572-015-0069-5

51   Johanna Hoerning, "Auf Definitionssuche: Die ‚Megastadt' Als Typus," in *Megastädte zwischen Begriff und Wirklichkeit: Über Raum, Planung und Alltag in großen Städten* (Bielefeld: transcript Verlag, 2016), 29–55.

52   Adey, "Air/Atmospheres," 294.

53   Celia Lowe, "Viral Clouds: Becoming H5N1 in Indonesia," *Cultural Anthropology* 25, no. 4 (November 2010): 625–49. https://doi.org/10.1111/j.1548-1360.2010.01072.x

54   Lowe, "Viral Clouds," 639.

55   Lowe, "Viral Clouds," 639–43.

56   Lowe, "Viral Clouds," 643.

57   Magdalena Górska, *Breathing Matters: Feminist Intersectional Politics of Vulnerability* (Linköping: Linköping University Electronic Press, 2016).

58   Górska, *Breathing Matters*, 94.

59 Gabriel O. Apata, "'I Can't Breathe': The Suffocating Nature of Racism," *Theory, Culture & Society* 37, no. 7–8 (December 2020): 241–54. https://doi.org/10.1177/0263276420957718

60 Apata, "'I Can't Breathe,'" 241–54.

61 Apata, "'I Can't Breathe,'" 243–46.

62 Möller, *Luft*, 63–69; Center for Science Education (UCAR), "Air Pollution: How We're Changing the Air," https://scied.ucar.edu/learning-zone/air-quality/air-pollution, accessed May 16, 2020.

63 Möller, *Luft*, 63–69; Aaron Daly and Paolo Zannetti, "An Introduction to Air Pollution—Definitions, Classifications, and History," in *Ambient Air Pollution*, ed. Paolo Zannetti, D. Al-Ajmi, and S. Al-Rasgied (The Arab School for Science and Technology, 2007), 1–14; Center for Science Education (UCAR), "Air Pollution."

64 Möller, *Luft*, 69.

65 Marc Z. Jacobson, *Air Pollution and Global Warming: History, Science, and Solutions* (Cambridge: Cambridge University Press, 2012), 73.

66 William M. Cavert, *The Smoke of London: Energy and Environment in the Early Modern City* (Cambridge: Cambridge University Press, 2016), viii–xx.

67 Timothy Morton, *Hyperobjects: Philosophy and Ecology after the End of the World* (Minneapolis: University of Minnesota Press, 2013).

68 Morton, *Hyperobjects*, 1.

69 Morton, *Hyperobjects*, 1–2.

70 Morton, *Hyperobjects*, 1–2; John R. Eperjesi, "Air Pollution in the Anthropocene: Fine Dust as Hyperobject," *Literature and Environment* 18, no. 2 (June 2019): 235–72. https://doi.org/10.36063/asle.2019.18.2.008

71 Morton, *Hyperobjects*, 1–2.

72 Purnamita Dasgupta and Kavitha Srikanth, "Reduced Air Pollution during COVID-19: Learnings for Sustainability from Indian Cities," *Global Transitions* 2 (2020): 271–82. https://doi.org/10.1016/j.glt.2020.10.002

73 Robert-Jan Wille, "Keep Focusing on the Air: COVID-19 and the Historical Value of an Atmospheric Sensibility," *Journal for the History of Environment and Society* 5 (January 2020): 181–93. https://doi.org/10.1484/J.JHES.5.122474; Khurram Shehzad, Muddassar Sarfraz, and Syed Ghulam Meran Shah, "The Impact of COVID-19 as a Necessary Evil on Air Pollution in India during the Lockdown," *Environmental Pollution* 266 (November 2020): 1–5. https://doi.org/10.1016/j.envpol.2020.115080

74 Mark Kinver, "Then and Now: Pandemic Clears the Air," BBC, June 1, 2021, https://www.bbc.com/news/science-environment-57149747, accessed May 17, 2021; Jonathan Watts, "Blue-Sky Thinking: How Cities Can Keep Air Clean after Coronavirus," *The Guardian*, June 7, 2020, https://www.theguardian.com/environment/2020/jun/07/blue-sky-thinking-how-cities-can-keep-air-clean-after-coronavirus, accessed June 3, 2022.

75 Kinver, "Then and Now: Pandemic Clears the Air."

76 Maarten Hajer and Wytske Versteeg, "Imagining the Post-Fossil City: Why Is It so Difficult to Think of New Possible Worlds?," *Territory, Politics, Governance* 7, no. 2 (April 3, 2019): 122–34. https://doi.org/10.1080/21622671.2018.1510339

77 Hajer and Versteeg, "Imagining the Post-Fossil City," 122–34.

78 Hajer and Versteeg, "Imagining the Post-Fossil City," 122–34.

79   Hajer and Versteeg, "Imagining the Post-Fossil City," 122–34.

80   Eddie Harmon-Jones and Judson Mills, "An Introduction to Cognitive Dissonance Theory and an Overview of Current Perspectives on the Theory," in *Cognitive Dissonance: Reexamining a Pivotal Theory in Psychology*, ed. Eddie Harmon-Jones (Washington, DC: American Psychological Association, 2019), 3.

81   Leon Festinger, *A Theory of Cognitive Dissonance* (Stanford, CA: Stanford University Press, 1957).

82   Festinger, *Cognitive Dissonance*, 1–32; Harmon-Jones and Mills, "Introduction to Cognitive Dissonance," 3–5.

83   Festinger, *Cognitive Dissonance*, 11.

84   Festinger, *Cognitive Dissonance*, 11.

85   Beth Gardiner, "Pollution Made COVID-19 Worse. Now, Lockdowns Are Clearing the Air.," National Geographic, April 8, 2020, https://www.nationalgeographic.com/science/article/pollution-made-the-pandemic-worse-but-lockdowns-clean-the-sky, accessed June 29, 2022; Zander S. Venter, Kristin Aunan, Sourangsu Chowdhury, and Jos Lelieveld, "COVID-19 Lockdowns Cause Global Air Pollution Declines," *Proceedings of the National Academy of Sciences* 117, no. 32 (August 11, 2020): 18986–90. https://doi.org/10.1073/pnas.2006853117

86   Venter et al., "COVID-19 Lockdowns," 18984–85.

87   Matthew A Cole, Ceren Ozgen, and Eric Strobl, "Air Pollution Exposure and COVID-19," *IZA Discussion Paper Series* 13367 (2020): 1–32. https://dx.doi.org/10.2139/ssrn.3628242; Venter et al., "COVID-19 Lockdowns," 18984–85. https://dx.doi.org/10.2139/ssrn.3628242

88   Xiao Wu, Rachel C. Nethery, M. Ben Sabath, Danielle Braun, and Francesca Domici, "Exposure to Air Pollution and COVID-19 Mortality in the United States," *ISEE Conference Abstracts* 2020, no. 1 (October 26, 2020), 1–6. https://doi.org/10.1289/isee.2020.virtual.O-OS-638

89   Wu et al., "Exposure to Air Pollution," 1–6.

90   Wu et al., "Exposure to Air Pollution," 1–6.

91   Anneleen Kenis and Maarten Loopmans, "Just Air? Spatial Injustice and the Politicisation of Air Pollution," *Environment and Planning C: Politics and Space* 40, no. 3 (April 11, 2022): 1–9. https://doi.org/10.1177/23996544221094144

92   Kenis and Loopmans, "Just Air?," 1–9.

93   Kenis and Loopmans, "Just Air?," 1–9.

94   Kenis and Loopmans, "Just Air?," 1–9.

95   Kenis and Loopmans, "Just Air?," 3–6.

96   Wade, "Livestock Drove Ancient Old World Inequality," 850.

97   Evan Hill, Ainara Tiefenthäler, Christiaan Triebert, Robin Stein, Drew Jordan, and Haley Willis, "How George Floyd Was Killed in Police Custody," *New York Times*, May 31, 2020, https://www.nytimes.com/2020/05/31/us/george-floyd-investigation.html, accessed July 16, 2022.

98   Dhaval Dave, Andrew Friedson, Kyutaro Matsuzawa, Joseph Sabia, and Samuel Safford, "Black Lives Matter Protests, Social Distancing, and COVID-19," *SSRN Electronic Journal* (2020): 1–61. https://doi.org/10.2139/ssrn.3631599

99   Amy Harmon and Rick Rojas, "A Delicate Balance: Weighing Protest Against the Risks of the Coronavirus," *New York Times*, June 7, 2020, https://www.nytimes.com/2020/06/07/us/Protest-coronavirus-george-floyd.html, accessed June 16, 2022.

100 Harmon and Rojas, "A Delicate Balance."

101 Elizabeth Day, "#BlackLivesMatter: The Birth of a New Civil Rights Movement,"
The Guardian, July 19, 2015, https://www.theguardian.com/world/2015/jul/19/
blacklivesmatter-birth-civil-rights-movement, accessed July 16, 2022.

102 Black Lives Matter, "About," https://blacklivesmatter.com/about/, accessed July 16,
2020.

103 Wolfgang Seibel, "Hegemoniale Semantiken und radikale Gegennarrative," *Universität
Konstanz*, January 22, 2009, 1–6.

104 Seibel, "Hegemoniale Semantiken," 1–6.

105 Seibel, "Hegemoniale Semantiken," 1–6.

106 Nikita Carney, "All Lives Matter, but so Does Race: Black Lives Matter and the
Evolving Role of Social Media," *Humanity & Society* 40, no. 2 (May 2016): 180–99.
https://doi.org/10.1177/0160597616643868

107 Scott Plous and Tyrone Williams, "Racial Stereotypes from the Days of American
Slavery: A Continuing Legacy," *Journal of Applied Social Psychology* 25, no. 9
(May 1995): 795–817. https://doi.org/10.1111/j.1559-1816.1995.tb01776.x

108 Michael Omi and Howard Winant, *Racial Formation in the United States* (New York:
Routledge/Taylor & Francis Group, 2015).

109 Omi and Winant, *Racial Formation*, 211–38.

110 Omi and Winant, *Racial Formation*, 217.

111 Omi and Winant, *Racial Formation*, 261; Carney, "All Lives Matter," 186.

112 Omi and Winant, *Racial Formation*, 220.

113 Omi and Winant, *Racial Formation*, 219.

114 Carney, "All Lives Matter," 186.

115 Arundhati Roy, "The Pandemic Is a Portal," in *Azadi: Freedom. Fascism. Fiction.*
(Chicago: Haymarket Books, 2020).

116 Gordon Fyfe and John Law, "Introduction: On the Invisibility of the Visual," *The
Sociological Review* 35, no. 1_suppl (May 1987): 1–14. https://doi.org/10.1111/j.1467-
954X.1987.tb00080.x

117 Luce Irigaray, *The Forgetting of Air in Martin Heidegger* (London: The Athlone Press,
1999).

118 Simone Dennis, "Explicating the Air: The New Smokefree (and Beyond)," *The
Australian Journal of Anthropology* 26, no. 2 (2015): 196–210. https://doi.org/10.1111/
taja.12103

Index

Note: Page numbers in *italics* denote figures.

Printed in the USA
CPSIA information can be obtained
at www.ICGtesting.com
JSHW021531051123
51465JS00002B/78

9 781804 131183